Elementary of Physical Chemistry

:: Author ::

Dr. Darshan V. Chaudhary

PUBLISHED BY

The New Era International Publishing House
H.Q. At & Po. Chaveli., Ta- Chansma,
Dist- Patan, North Gujarat, India, Asia.
www.iphouseindia.com

First Publication: 15th August, 2015

ISBN:- 978-1-51722-013-6

Price: Rs.800/- INDIA

$ 10 OUTSIDE INDIA

PUBLISHED BY

The New Era International Publishing House

H.Q. At & Po. Chaveli., Ta- Chansma,

Dist- Patan, North Gujarat, India, Asia.

www.iphouseindia.com

PREFACE

Where has the Physical Chemistry come from? Throughout the antiquity of humanoid competition, societies have thrashed to make intelligence of the world around them. Through the division of science we call Chemistry we have increased an understanding of the chemical bonding in different atoms, structures and molecules, different colligative properties which makes up our world and of the interactions between chemical properties on which it depends. Important advances in our understanding of the nature of physical chemistry of different elememts and their action were made in the late 18th and 19th centuries, seeding the explosive expansion from the 1850s and 60s onward to the present billion dollar physical chemistry based industries.

Much of the information about different atoms and molecules have been made possible because they have functions largely because of their mode of action. Life is a dynamic process that involves constant changes in chemical composition. These changes are regulated by catalytic reactions, their physical and chemical properties which are regulated by biological activities involving wide across the surrounding. Few chemists and allied biologists continue to think of this as a simple task, but we know that environment as we know it could not exist or a ailment would not be cured or prevented successfully without the use of targeted specific atoms, molecules and elements whilst a proper due medication.

Recent developments in the fields of physical chemistry and chemical kinetic, equilibrium, bonding are bringing ever more powerful means of analysis to bear on the study of atoms, molecules or elements structure and function that will undoubtedly leading to the rational modifications to match specific requirements and also the design of new atoms, molecules or elements with novel and multifunctional or a broad spectrum properties.

In present book, chapter 1 introduces the chemical bonding in different atoms, molecules and elements of selected different bonds. Chapter 2 describes the various colligative properties study. Chapter 3 elobarates different thermodynamic processes on, while chapter 4 emphasis chemical equilibrium, Chapter 5 discusses chemical kinetics, Chapter 6 elaborates scope, the solid state chemistry and the last chapter 7 emphases on gaseous state of different atoms, elements and molecules.

The contents of the book will be useful to the students of Chemistry, Biotechnology, Industrial Chemists, Pharmaceutical science and technology, Physical Chemistry, Public health sciences etc.

I express my heartfelt thanks to Dr. Pranav Srivastav, Professor, Chemistry Department, Gujarat University, Ahmedabad, India, for his constant guidance across my research Work and without the same platform I would not be able to compile this book. I am very thankful to my other colleague contributor Mr. Edvin Pithawala for critical evaluation of each chapter across the book. I would like to express my gratitude to my family members especially to my parents for their love affection and care and last but not the least to my beloved *Reema* for her everlasting love, motivation and sacrifice for the time taken in compiling this book.

I am grateful to publisher for their concern, efforts and encouragement, especially for their excellent cooperation in the task of preparing and publishing this book.

- **Dr. Darshan Chaudhary**

TABLE OF CONTENTS

CHAPTER – 1

CHEMICAL BONDING

OBJECTIVES

After Studying this Chapter you will able to:

- *To know about bonding as binding forces between atoms to form molecules.*
- *To learn about Kossel-Lewis approach to chemical bonding, the octet rule, its limitations and Lewis representations of simple molecules.*
- *To know about ionic bond, lattice energy and Born-Haber cycle.*
- *To understand covalent bond, directional character.*
- *To learn about VSEPR model and predict the geometry of simple molecules.*
- *To understand the concepts of hybridization,* 1DQGŒERQGV UHVRQDQFH *and coordinate covalent bonds.*

1.1 Elementary theories on Chemical Bonding

The study on the "nature of forces that hold or bind atoms together to form a molecule" is required to gain knowledge of the following-

i) to know about how atoms of same element form different compounds combining with different elements.

ii) to know why particular shapes are adopted by molecules.

iii) to understand the specific properties of molecules or ions and the relation between the specific type of bonding in the molecules.

Chemical bond

Existence of a strong force of binding between two or many atoms is referred to as a **Chemical Bond** and it results in the formation of a stable compound with properties of its own. The bonding is permanent until it is acted upon by external factors like chemicals, temperature, energy etc. It is known that, a molecule is made up of two or many atoms having its own characteristic properties which depend on the types of bonding present.

Classification of molecules

Molecules having two identical atoms like H_2, O_2, Cl_2, N_2 etc. are called as **homonuclear diatomic molecules**. Molecules containing two different atoms like CO, HCl, NO, HBr etc., are called as **heteronucleardiatomic molecules.** Molecules containing identical but many atomsbonded together such as P_4, S_8 etc., are called as **homonuclearpolyatomics.** In most of the molecules, more than two atoms of differentkinds are bonded such as in molecules like NH_3, CH_3COOH, SO_2, HCHO and they are called as **heteronuclearpolyatomics.**

Chemical bonds are basically classified into three types consisting of (i) ionic or electrovalent bond (ii) covalent bond and (iii) coordinate-covalent bond. Mostly, valence electrons in the outer energy level of an atom take part in the chemical bonding.

In 1916, W.Kossel and G.N.Lewis, separately developed theories of chemical bonding inorder to understand why atoms combined to form molecules. According to the electronic theory of valence, a chemical bond is said to be formed when atoms interact by losing, gaining or sharing of valence electrons and in doing so, a stable noble gas electronic configuration is achieved by the atoms.

Except Helium, each noble gas has a stable valence shell of eight electrons. The tendency for atoms to have eight electrons in their outershell by interacting with other atoms through electron sharing or electron-transfer is known as the **octet rule** of chemical bonding.

1.1.1 Kossel-Lewis approach to Chemical Bonding

W.Kossel laid down the following postulates to the understanding of ionic bonding:

- In the periodic table, the highly electronegative halogens and the highly electropositive alkali metals are separated by the noble gases. Therefore one or small number of electrons are easily gained and transferred to attain the stable noble gas configuration.

The formation of a negative ion from a halogen atom and a positive ion from an alkali metal atom is associated with the gain and loss of an electron by the respective atoms.

- The negative and positive ions so formed attains stable noble gas electronic configurations. The noble gases (with the exception of helium which has two electrons in the outermost shell) have filled outer shell electronic configuration of eight electrons (octet of electrons) with a general representation ns^2 np^6.

- The negative and positive ions are bonded and stabilised by force of electrostatic attraction.

Kossel's postulates provide the basis for the modern concepts on electron transfer between atoms which results in ionic or electrovalent bonding.

For example, formation of NaCl molecule from sodium and chlorine atoms can be considered to take place according to Kossel's theory by an electron transfer as:

(i) $Na \rightarrow Na^+ + e^-$

[Ne] $3s^1$ [Ne] where [Ne] = electronic configuration of Neon $= 2s^2\ 2p^6$

(ii) $Cl + e \rightarrow Cl^-$

[Ne] $3s^2\ 3p^5$ [Ar]

[Ar] = electronic configuration of Argon

(iii) $Na^+ + Cl^- \rightarrow NaCl$ (or) Na^+Cl^-

NaCl is an electrovalent or ionic compound made up of sodium ions and chloride ions. The bonding in NaCl is termed as electrovalent or ionic bonding. Sodium atom loses an electron to attain Neon configuration and also attains a positive charge. Chlorine atom receives the electron to attain the Argon configuration and also becomes a negatively charged ion. The columbic or electrostatic attraction between Na^+ and Cl^- ions result in NaCl formation.

Similarly formation of MgO may be shown to occur by the transfer of two electrons as:

(i) $Mg \rightarrow Mg^{+2} + 2e^-$

[Ne] $3s^2$ [Ne]

(ii) $O + 2e \rightarrow O^{2-}$

[He]$2s^2$ $2p^4$ [He]$2s^2$ $2p^6$ (or) [Ne]

(iii) Mg^{+2} + O^{2-}→MgO (or) Mg^{+2} O^{2-}

The bonding in MgO is also electrovalent or ionic and the electrostatic forces of attraction binds Mg^{2+} ions with O^{2-} ions. Thus, "the binding forces existing as a result of electrostatic attraction between the positive and negative ions", is termed as **electrovalent** or **ionic** bond. The electrovalency is considered as equal to the number of charges on an ion. Thus magnesium has positive electrovalency of two while chlorine has negative electrovalency of one.

The valence electron transfer theory could not explain the bonding in molecules like H_2, O_2, Cl_2 etc., and in other organic molecules that have ions.

G.N.Lewis, proposed the octet rule to explain the valence electron sharing between atoms that resulted in a bonding type with the atoms attaining noble gas electronic configuration. The statement is: "a bond is formed between two atoms by mutual sharing of pairs of electrons to attain a stable outer-octet of electrons for each atom involved in bonding". This type of valence electron sharing between atoms is termed as **covalentbonding.** Generally homonucleardiatomics possess covalent bonds.

It is assumed that the atom consists of a `Kernel' which is made up of a nucleus plus the inner shell electrons. The Kernel is enveloped by the outer shells that could accommodate a maximum of eight electrons. The eight outershell electrons are termed as octet of electrons and represents a stable electronic configuration. Atoms achieve the stable outer octet when they are involved in chemical bonding.

In case of molecules like F_2, Cl_2, H_2 etc., the bond is formed by the sharing of a pair of electrons between the atoms. For example, consider the formation of a fluorine molecule (F_2). The atom has electronic configuration. $[He]2s^2 3s^2 3p^5$ which is having one electron less than the electronic configuration of Neon. In the fluorine molecule, each atom contributes one electron to the shared pair of the bond of the F_2 molecule. In this process, both the fluorine atoms attain the outershell octet of a noble gas (Argon) (Fig. 10.1(a)). Dots (•) represent electrons. Such structures are called as Lewis dot structures.

Lewis dot structures can be written for combining of like or different atoms following the conditions mentioned below :

- Each bond is the result of sharing of an electron pair between the atoms comprising the bond.
- Each combining atom contributes one electron to the shared pair.
- The combining atoms attain the outer filled shells of the noble gas configuration.

If the two atoms share a pair of electrons, a single bond is said to be formed and if two pairs of electrons are shared a double bond is said to be formed etc. All the bonds formed from sharing of electrons are called as covalent bonds.

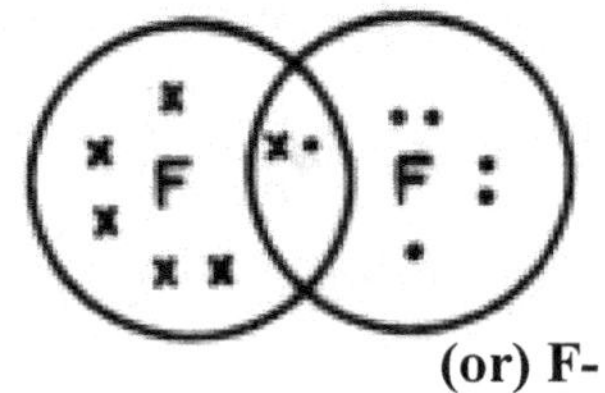

F 8e⁻ 8e⁻

Fig. 1.1(a) F_2 molecule

In carbon dioxide (CO_2) two double bonds are seen at the centre carbon atom which is linked to each oxygen atom by a double bond. The carbon and the two oxygen atoms attain the Neon electronic configuration.

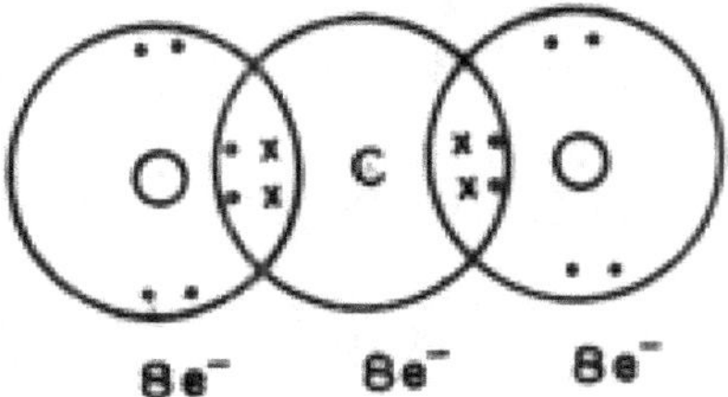

Fig. 1.1(b) CO_2 molecule

When the two combining atoms share three electron pairs as in N_2 molecule, a triple bond is said to be formed. Each of the Nitrogen atom shares 3 pairs of electrons to attain neon gas electronic configuration.

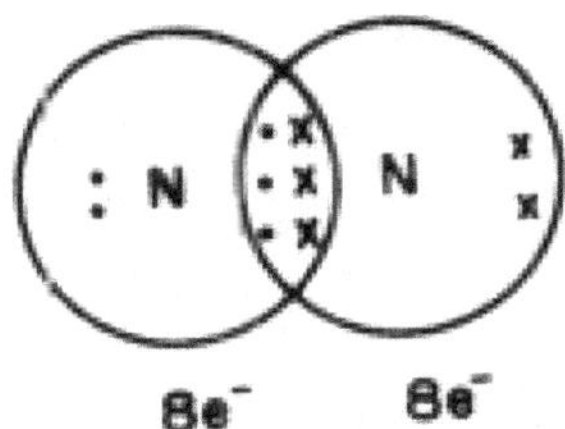

Fig. 1.1 (c) N_2 molecule

1.2 Types of Bond

There are more than one type of chemical bonding possible between atoms which makes the molecules to show different characteristic properties. The different types of chemical bonding that are considered to exist in molecules are (i) **ionic or electrovalent bond** which is formed as a result of complete electron transfer from one atom to the other that constitutes the bond; (ii) **covalent bond** which is formed as a result of mutual electron pair sharing with an electron being contributed by each atom of the bond and (iii)

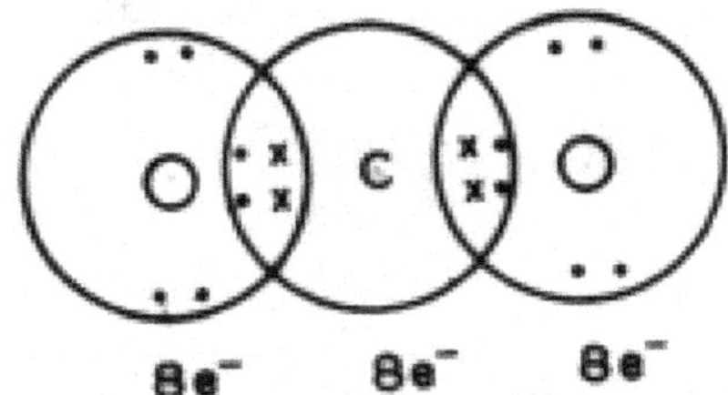

Co-ordinate-covalent bond which is formed as a result of electron pair sharing with the pair of electrons being donated by only one atom of the bond. The formation and properties of these types of bonds are discussed in detail in the following sections.

1.3 Ionic (or) Electrovalent bond

The electrostatic attraction force existing between the cation and the anion produced by the electron transfer from one atom to the other is known as the ionic (or) electrovalent bond. The compounds containing such a bond are referred to as ionic (or) electrovalent compounds.

Ionic bond is non directional and extends in all directions. Therefore, in solid state single ionic molecules do not exist as such. Only a network of cations and anions which are tightly held together by electro-static forces exist in the ionic solids. To form a stable ionic compound there must be a net lowering of energy. That is, energy is released as a result of electrovalent bond formation between positive and negative ions.

When the electronegativity difference between the interacting atoms are greatly different they will form an ionic bond. In fact, a difference of 2 or more is necessary for the formation of an ionic bond. Na has electronegativity 0.9 while Cl has 3.0, thus Na and Cl atoms when brought together will form an ionic bond.

For example, NaCl is formed by the electron ionisation of sodium atom to Na^+ ion due to its low ionisation potential value and chlorine atom to chloride ion by capturing the odd electron due to high electron affinity. Thus, NaCl (ionic compound) is formed. In NaCl, both the atoms possess unit charges.

i) $Na_{(g)} \rightarrow Na^{-}_{(g)} + e^-$

$2s^2 2p^6 3s^1$ $\quad 2s^2 2p^6$ sodium cation

ii) $Cl_{(g)} + e^- \rightarrow Cl^-$Chloride ion

$3s^2 3p^5$ $\quad 3s^2, 3p^5$

iii) $Na^+ + Cl^- \rightarrow NaCl$

Sodium ion Chloride ionionic/crystalline compound is formed

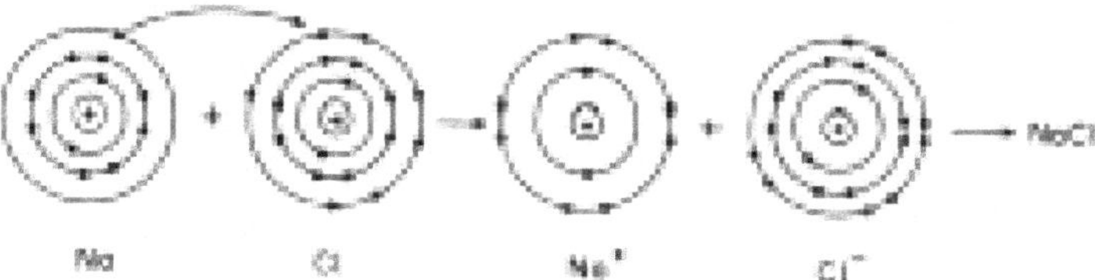

Fig. 1.2 Electron transfer between Na and Cl atoms during ionic bond formation in NaCl

In CaO, which is an ionic compound, the formation of the ionic bond involves two electron transfers from Ca to O atoms. Thus, doubly charged positive and negative ions are formed.

$Ca \xrightarrow{\text{ionisation}} Ca^{2+} + 2e^-$ (Calcium Cation) $3p^6\ 4s^2$

$O + 2e \xrightarrow{\text{electron}} O^{2-}$(Oxide anion)

$2s^2 2p^4$ $\quad 2s^2 2p^6$

$Ca^{2+} + O^{2-} \longrightarrow CaO$

Ionic compound

Ionic bond may be also formed between a doubly charged positive ion with single negatively charged ion and vice versa. The molecule as a whole remains electrically neutral. For example in MgF_2, Mg has two positive charges and each fluorine atom has a single negative charge. Hence, Mg^{2+} binds with two fluoride (F^-) ions to form MgF_2 which is electrically neutral.

$Mg \rightarrow \quad Mg^{+2} \quad + \quad 2e^-$

$2s^2 2p^6 3s^2 \quad 2s^2\, 2p^6$

$2e \quad + \quad 2F \quad \rightarrow \quad 2F^-$

$2s^2\, 2p^5 \qquad\qquad 2s^2\, 2p^6$

i.e.: $Mg^{+2} + 2F^- \rightarrow MgF_2$

Magnesium – fluoride (an ionic compound)

Similarly in Aluminum bromide ($AlBr_3$), Aluminum ion has three positive charges and therefore it bonds with three Bromide ions to form $AlBr_3$ which is a neutral ionic molecule.

$Al \quad \rightarrow \quad Al^{3+} \quad + \quad 3e^-$

$2p^6 3s^2 3p^1 \quad 2s^2 2p^6$

$3\,Br + 3e^- \rightarrow 3Br^-$

$4s^2\, 4p^5 \qquad\qquad 4s^2\, 4p^6$

$Al^{3+} + 3Br^- \rightarrow AlBr_3$ (ionic bond)

1.3.1 Lattice energy and Born - Haber's cycle

Ionic compounds in the crystalline state exist as three dimensionally ordered arrangement of cations and anions which are held together by columbic interaction energies. The three dimensional network of points that represents the basic repetitive arrangement of atoms in a crystal is known as lattice or a space lattice. Thus a qualitative measure of the stability of an ionic compound is provided by its enthalpy of lattice formation.

Lattice enthalpy of an ionic solid is defined as the energy required to completely separate one mole of a solid ionic compound into gaseous constituent ions. That is, the enthalpy change of dissociation of MX ionic solid into its respective ions at infinity separation is taken the lattice enthalpy.

$$MX_{(s)}\ M^{+}_{(g)} + X^{-}_{(g)}$$

$$\Delta_r H^{\circ} = L.E$$

Lattice enthalpy is a positive value.

For example, the lattice enthaply of NaCl is 788 $kJ.mol^{-1}$. This means that 788 kJ of energy is required to separate 1 mole of solid NaCl into 1 mole of $Na^{+}_{(g)}$ and 1 mole of $Cl^{-}_{(g)}$ to an infinite distance.

In ionic solids, the sum of the electron gain enthalpy and the ionization enthalpy may be positive but due to the high energy released in the formation of crystal lattice, the crystal structure gets stabilized.

Born Haber's Cycle Determination of Lattice enthalpy

It is not possible to calculate the lattice enthalpy directly from the forces of attraction and repulsion between ions but factors associated with crystal geometry must also be included. The solid crystal is a three-dimensional entity. The lattice enthalpy is indirectly determined by the use of Born - Haber Cycle. The procedure is based on Hess's law, which states that the enthalpy change of a reaction is the same at constant volume and pressure whether it takes place in a single or multiple steps long as the initial reactants and the final products remain the same. Also it is assumed that the formation of an ionic compound may occur either by direct combination of elements (or) by a step wise process involving vaporization of elements, conversion of gaseous atoms into ions and the combination of the gaseous ions to form the ionic solid.

For example consider the formation of a simple ionic solid such as an alkali metal halide MX, the following steps are considered.

$$M_{(s)} \xrightarrow{\Delta H^0_{(1)}} M_{(g)} \xrightarrow{\Delta H^0_{(3)}} M^+_{(g)} + e$$

$$+ \qquad\qquad\qquad\qquad +$$

$$1/2X_{2(g)} \xrightarrow{\Delta H^0 2} X_{(g)} \xrightarrow[+e]{\Delta H^0 4} X^-_{(g)}$$

$$\Delta H^0_f \qquad \downarrow$$

$$MX_{(s)}$$

û+°1 = enthalpy change for sublimation of $M_{(s)}$ to $M_{(g)}$

û+°$_2$ = enthalpy change for dissociation of 1/2 $X_{2(g)}$ to $X_{(g)}$

û+°$_3$= ionization energy of $M_{(g)}$ to $M^+_{(g)}$

û+°$_4$= electronic affinity or electron gain energy for conversion of $X_{(g)}$ to $X^-_{(g)}$

û+°$_5$ = the lattice enthalpy for formation of solid MX (1 mole).

û$_f$H°= enthalpy change for formation of MX solid directly from the respective elements such as 1 mole of solid M and 0.5 moles of $X_{2(g)}$.

According to Hess's law,

û+°f = û+°1 û+°$_2$ û+°$_3$ û+°$_4$ û+°$_5$*

Some important features of lattice enthalpy are:

i. The greater the lattice enthalpy the more stable the ionic bond formed.

ii. The lattice enthalpy is greater for ions of higher charge and smaller radii.

iii. The lattice enthalpies affect the solubilities of ionic compounds.

Calculation of lattice enthalpy of NaCl

Let us use the Born-Haber cycle for determining the lattice enthalpy of NaCl as follows :

7KH VWDQGDUG HQWKDOS\ FKDQJH û$_f$H° overall for the reaction, $Na_{(s)} + 1/2Cl_{2(g)} \rightarrow NaCl_{(s)}$ is - 411.3 $kJmol^{-1}$

$$Na_{(s)} + \tfrac{1}{2} Cl_{2(g)} \xrightarrow{\Delta_f H^o} NaCl_{(s)}$$

* 7KH YDOXH RI û+°$_5$ calculated using the equation of Born-Haber cycle should be reversed in sign

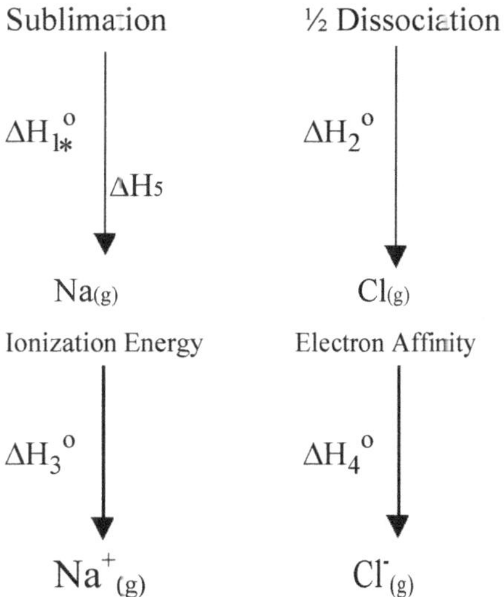

Fig. 1.3 Born-Haber cycle for Lattice enthalpy determination involving various stepwise enthalpic processes for NaCl solid formation

Since the reaction is carried out with reactants in elemental forms and products in their standard states, at 1 bar, the overall enthalpy change of the reaction is also the enthalpy of formation for NaCl. Also, the formation of NaCl can be considered in 5 steps.Thesum of the enthalpy changes of these steps is considered equal to the enthalpy change for the overall reaction from which the lattice enthalpy of NaCl is calculated.

Atomization :

û+°$_1$ for $Na_{(s)} \rightarrow Na_{(g)}$ is + 108.70 (kJ mol[1]) Dissociation:

û+°$_2$ for ½ $Cl_{2(g)} \rightarrow Cl_{(g)}$ is + 122.0

Ionization :

û+°$_3$ for $Na_{(g)} \rightarrow Na^+_{(g)}$ + e is + 495.0 Electron affinity :

û+°$_4$ for e + $Cl_{(g)} \rightarrow Cl^-_{(g)}$ is - 349.0 Lattice enthalpy

û+°$_5$ for $Na^+_{(g)} + Cl^-_{(g)} \rightarrow NaCl_{(g)}$ is ?

û$_f$H° û+°$_1$ û+°$_2$ û+°$_3$ û+°$_4$ û+°$_5$

-411.3 = 108.70 + 122.0 + 495 -û+°$_5$

∴ û+°$_5$ = -788.0 kJ mol^{-1}

But the lattice enthalpy of NaCl is defined by the reaction $NaCl_{(g)} \rightarrow Na^+_{(g)} + Cl^-_{(g)}$ only.

∴ /DWWLFH HQWKDOS\ YDOXH IURP û+°5is written with a reversed sign.

∴Lattice enthalpy of NaCl = +788.0 kJ mol^{-1}.

Ionic compounds possess characteristic properties of their own like physical state, solubility, melting point, boiling point and conductivity. The nature of these properties are discussed as follows.

i. Due to strong columbic forces of attraction between the oppositely charged ions, electrovalent compounds exist mostly as hard crystalline solids. Due to the hardness and high lattice enthalpy, low volatility, high melting and boiling points are seen.

ii. Because of the strong electrostatic forces, the ions in the solid are not free to move and act as poor conductor of electricity in the solid state. However, in the molten state, or in solution, due to the mobility of the ions electrovalent compounds become good conductor of electricity.

iii. Ionic compounds possess characteristic lattice enthalpies since they exist only as ions packed in a definite three dimensional manner. They do not exist as single neutral molecule or ion.

iv. Ionic compounds are considered as polar and are therefore, soluble in high dielectric constant solvents like water. In solution, due to solvation of ions by the solvent molecules, the strong interionic attractions are weakened and exist as separated ions.

v. Electrovalent compounds having the same electronic configuration exhibit isomorphism.

1.4 Covalent bond

A covalent bond is a chemical bond formed when two atoms mutually share a pair of electron. By doing so, the atoms attain stable octet electronic configuration. In covalent bonding, overlapping of the atomic orbitals having an electron from each of the two atoms of the bond takes place resulting in equal sharing of the pair of electrons. Also the interatomic bond thus formed due to the overlap of atomic orbitals of electrons is known as a covalent bond. Generally the orbitals of the electrons in the valency shell of the atoms are used for electron sharing. The shared pair of electrons lie in the middle of the covalent bond. Including the shared pair of electrons the atoms of the covalent bond attain the stable octet configuration. Thus in hydrogen molecule (H_2) a covalent bond results by the overlap of the two s orbitals each containing an electron from each of the two H atoms of the molecule. Each H atom attains '$1s^2$' filled K shell.

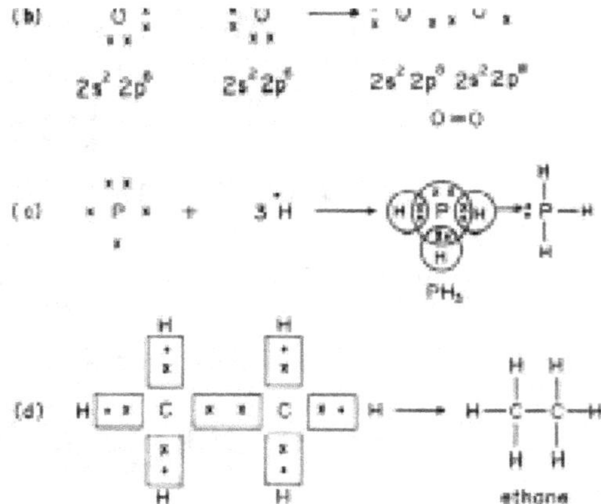

Fig. 1.4 Lewis dot structures of (a) Cl_2 (b) O_2 (c) PH_3 and (d) ethane molecules

Double bond formation

In oxygen (O_2) molecule, two pairs of electrons are mutually shared and a double bond results. The electronic configuration of O atom is $1s^2\ 2s^2\ 2p^4$. By sharing two more electrons from the other O atom, each O atom attains $2s^2\ 2p^6$, filled configuration. Thus O_2 molecule is represented as O=O. Similar to oxygen molecule in ethylene which is an organic molecule, a double covalent bond exists between the two carbon atoms due to the mutual sharing of two pairs of electrons. Each carbon atom attains the stable octet electron configuration.

1.4.1 Characteristics of covalent compounds

1. Covalent compounds are formed by the mutual sharing of electrons. There is no transfer of electrons from one atom to another and therefore no charges are created on the atom. No ions are formed. These compounds exist as neutral molecules and not as ions. Although some of the covalent molecules exist as solids, they do not conduct electricity in fused or molten or dissolved state.
2. They possess low melting and boiling points. This is because of the weak intermolecular forces existing between the covalent molecules. Since, no strong columbic forces are seen, some of covalent molecules are volatile in nature. Mostly covalent compounds possess low melting and boiling points.
3. Covalent bonds are rigid and directional therefore different shapes of covalent molecules are seen.
4. Most of the covalent molecules are non polar and are soluble in nonpolar (low dielectric constant) solvents like benzene, ether etc. and insoluble in polar solvents like water. Carbon tetrachloride (CCl_4) is a covalent nonpolar molecule and is soluble in benzene.

1.4.2 Fajan's rules

Covalent character of ionic bonds

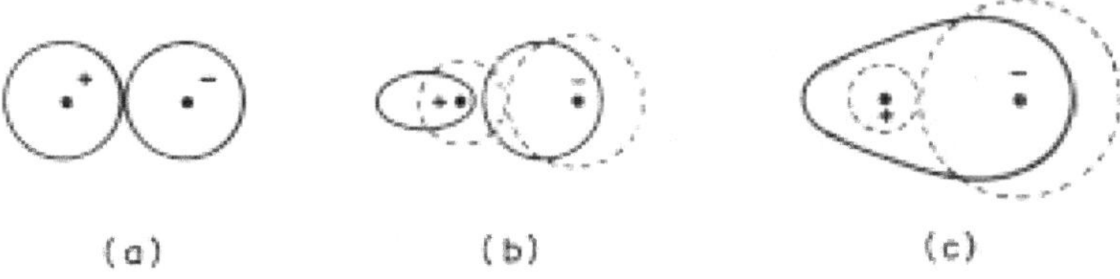

Fig. 1.5 Polarization effects : (a) idealized ion pair with no polarization, (b) mutually polarized ion pair (c) polarization sufficient to form covalent bond. Dashed lines represent hypothetical unpolarized ions

When cations and anions approach each other, the valence shell of anions are pulled towards cation nucleus due to the columbic attraction and thus shape of the anion is deformed. This phenomenon of deformation of anion by a cation is known as **polarization** and the ability of cation to polarize a nearby anion is called as polarizing power of cation.

Fajan points out that greater is the polarization of anion in a molecule, more is covalent character in it. This is **Fajan's rule.**

Fajan also pointed out the influence of various factors on cations for polarization of anion.

(i) When the size of a cation is smaller than a cation with the same charge, then the smaller sized cation causes a greater extent of polarization on the anion than the larger sized cation.

(ii) The polarizing capacity of a cation is related to its ionic potential (which is Z^+/r) which is inversely related to the ionic radius. Therefore comparing Li^+ and Na^+ or K^+ ions, although these cations have single positive charge, Li^+ ion polarizes an anion more than Na^+ or K^+ ions can do on the same anion. This is because of the smaller size of Li^+ than Na^+ or K^+ ions.

(iii) Greater the polarization effects greater will be the covalent character imparted into the ionic bond.

The general trend in the polarizing power of cations: $Li^+ > Na^+ > K^+ > Rb^+ > Cs^+$

covalent character:

$$LiCl > NaCl > KCl > RbCl > CsCl.$$

a) Size of the anion

When the size of anion is larger, valence electrons are less tightly held by its nucleus. Therefore more effectively the cation pulls the valence electrons towards its nucleus. This results in more polarization effect. That is, for the same charge of the anion, larger sized anion is more polarized than a smaller sized anion.

The trend in the polarization of anions:

$$I^- > Br^- > Cl^- > F^-$$

∴covalent character :

$$LiF < LiCl < LiBr < LiI$$

b) Charge on cation

If the oxidation state of the cation is higher the polarization of anion will be more. Thus more will be covalent nature in the bonding of the molecule.

Thus polarizing power: $Fe^{+2} < Fe^{+3}$

∴ Covalent character : $FeCl_2 < FeCl_3$.

c) Presence of polar medium

Presence of a polar medium keeps away the cations and anions from each other due to solvation. This prevents polarization of anion by the cation. Therefore $AlCl_3$ behaves as an ionic molecule in water, while it is a covalent molecule in the free state.

1.4.3 Polarity of Covalent Bonds

The existence of a purely ionic or covalent bond represents an ideal situation. In the covalently bonded molecules like H_2, Cl_2,

F_2(homonucleardiatomics), the bond is a pure covalent bond. In case of heteronuclear molecules like, HF, HCl, CO, NO etc, the shared electron pair gets displaced more towards the atom possessing higher electronegativity value than the other one. In HF, the shared electron pair is displaced more towards fluorine because the electronegativity of Fluorine is far greater than that of Hydrogen. This results in partial ionic character induced in the covalent bond and is represented as:

/ /-

H - F

However, no specific charges are being found on H or F and the molecule as a whole is neutral. Thus the extent of ionic character in a covalent bond will depend on the relative attraction of electrons of the bonded atoms which depends on the electro negativity differences between the two atoms constituting the covalent bond.

As a result of polarization, the molecule possessed a dipole moment. In a triatomic molecule like water two covalent bonds exist between the oxygen atom and the two H atoms. Oxygen with higher electronegativity attracts the shared pair of electrons to itself and thus oxygen becomes the negative end of the dipole while the two hydrogen atoms form the positive end. Thus the two covalent bonds in the water molecule possess partial ionic character.

Generally larger the electronegativity difference between the atoms consisting the bond, greater will be the ionic character. For H atom electronegativity is 2.1 and for Cl atom it is 3.0. Thus H-Cl covalent bond is polarized and it has more ionic character.

/ /-H –Cl

Consider the molecule like hydrogen cyanide HCN, the bond between hydrogen atoms and the cyanide anion is of covalent type. CN^- ion has more capacity to pull the shared pair of electrons in the H-CN bond that, partially H^+ CN^- are created. Thus in water medium this compound is ionised into H^+ and CN^- ions.

$$/ /\text{-H - C} \equiv \text{N} \rightarrow \text{H - CN}$$

1.5 Valence Shell Electron Pair Repulsion Theory (VSEPR) Theory

Molecules exist in different shapes. Many of the physical and chemical properties of molecules arise due to different shapes of the molecules.

Some of the common geometrical shapes found among the molecules are: linear, trigonal, planar, tetrahedral, square planar, trigonal-bipyramidal, square-pyramidal, octahedral, pentagonal-bipyramidal etc. The VSEPR theory provides a simple treatment for predicting the shapes of polyatomic molecules. The theory was originally proposed by Sigdwick and Powell in 1940. It was further developed and modified by Nyholm and Gillespie (1957).

The basic assumptions of the VSEPR theory are that:

i) Pairs of electrons in the valence shell of a central atom repel each other.

ii) These pairs of electrons tend to occupy positions in space that minimize repulsions and maximize the distance of separation between them.

iii) The valence shell is taken as a sphere with electron pairs localizing on the spherical surface at maximum distance from one another.

iv) A multiple bond is treated as if it is a single electron pair and the two or three electron pairs of a multiple bond are treated as a single super pair.

v) Where two or more resonance structures can depict a molecule the VSEPR model is applicable to any such structure.

It is convenient to divide molecule into two categories (i) molecules in which the central atom has no lone pairs of electrons and (i) molecules in which the central atom has one or more lone pairs.

Table 1.1 shows the different geometries of molecules or ions with central atom having no lone pair of electrons and represented by general type AB_x. In compounds of AB_2, AB_3, AB_4, AB_5, AB_6, types the arrangement of electron pairs (bonded pairs) as well as the B atoms around the central atom A are, linear, trigonal planar, tetrahedral, trigonal-bipyramidal and octahedral respectively. Such arrangements are present in $BeCl_2$ (AB_2); BF_3 (AB_3); CH_4 (AB_4) and PCl_5 (AB_5) molecules with geometries as shown below in Fig.1.6.

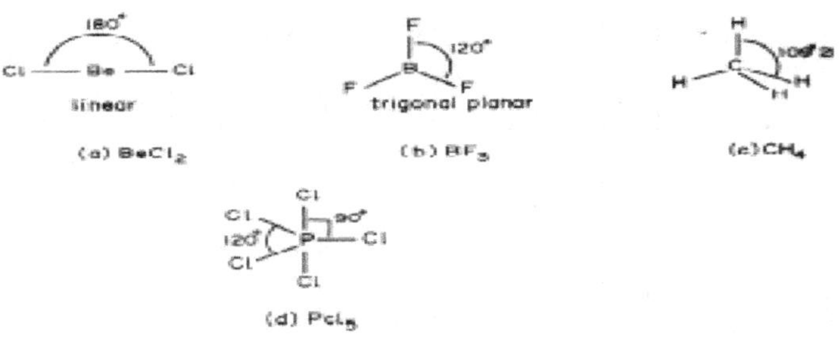

Fig. 1.6 Geometrical structures of some molecules (a) $BeCl_2$(b) BF_3 (c) CH_4 and (d) PCl_5

Table 1.1Geometry of molecules in which the central atomhas no lone pair of electrons

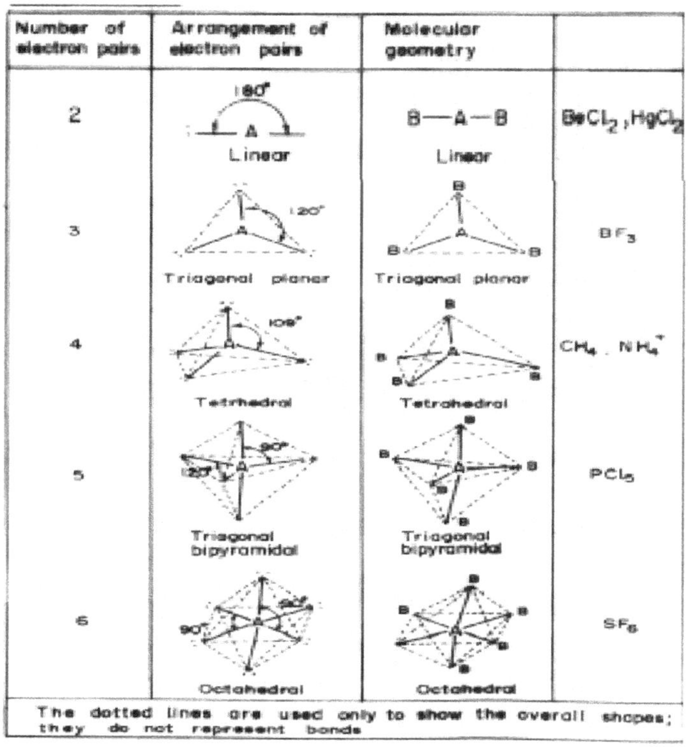

Number of electron pairs	Arrangement of electron pairs	Molecular geometry	
2	180° Linear	B—A—B Linear	$BeCl_2$, $HgCl_2$
3	120° Triagonal planer	Triagonal planar	BF_3
4	109° Tetrahedral	Tetrahedral	CH_4, NH_4^+
5	90° 120° Triagonal bipyramidal	Triagonal bipyramidal	PCl_5
6	90° 90° Octahedral	Octahedral	SF_6

The dotted lines are used only to show the overall shapes; they do not represent bonds

In case of molecules with the central atom having one or more lone pairs VSEPR treatment is as follows: In these type of molecules, both lone pairs and bond pairs of electrons are present. The lone pairs are localised on the central atom, and bonded pairs are shared between two atoms. Consequently, the lone pair electrons in a molecule occupy more space as compared to the bonding pair electrons. This causes greater repulsions between lone pairs of electrons as compared to the lone pairs of electrons to the lone pair (lp) - bonding pair and bonding pair - bonding pair repulsions (bp).

The descending order of repulsion interaction is

lp - lp>lp - bp>bp – bp

These repulsion effects cause deviations from idealized shapes and alterations in the predicted bond angles in molecules.

Table 10.2

Molecule Type	No of Bonding Pairs	No. of lone pairs	Arrangement of electron pairs	Shape (Geometry)	Examples
AB_3E	2	1	B–A–B Trigonal planar	Bent	SO_2, O_3
AB_3E	3	1	B–A(–B)–B Tetrahedral	Trigonal pyramidal	NH_3
AB_2E_2	2	2	:–A–B, B	Bent	H_2O

Examples: In sulphur dioxide molecule there are three electron pairs on the S atom. The overall arrangement is trigonal planar. However, because one of the three electron pairs is a lone pair, the SO_2 molecule has a `bent' shape and due to the lp - lp repulsive interactions the bond angle is reduced to 119.5° from the value of 120°.

In the ammonia (NH_3) molecule, there are three bonding pairs and one lone pair of electrons. The overall arrangement of four electron pairs is tetrahedral. In NH_3, one of the electron pairs, on nitrogen atom is a lone pair, so the geometry of NH_3 is pyramidal (with the N atom at the apex of the pyramid). The three N-H bonding pairs are pushed closer because of the lp-bp repulsion and the HNH angle gets reduced from 109°28' (which is the tetrahedral angle) to 107°.

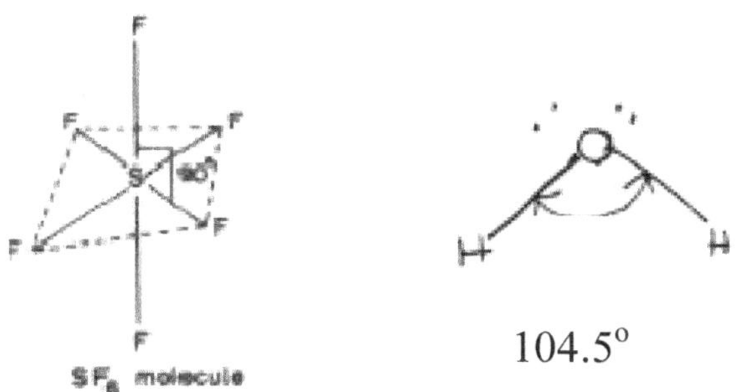

SF_6 molecule

104.5°

The water H_2O molecule, oxygen atom contains two bonding pairs and two lone pairs of electrons. The overall arrangement for four electron pairs is tetrahedral, but the lp-lp repulsions being greater than lp-bp repulsions in H_2O. The HOH angle is reduced to 104.5° than 109°28'. The molecule has a bent shape.

The molecule SF_6 belongs to AB_6 type consisting of 6 bp of electrons around the central sulphur atom. The geometrical arrangement will be a regular octahedral.

1.6 Directional Properties of Covalent Bonds

When the overlapping of orbitals occur along the internuclear axis (Line joining the two nuclei) then the electron orbitals merge to form cylindrically symmetrical regioQ DQG WKH ERQG LV FDOOHG DV C1 ERQG ,Q D 1 bond, maximum extent overlap of orbitals are possible and the bond formed is also stronger.

For e.g : H^-+ ERQG LV D 1 ERQG

Consider the valence bond description of O_2molecule: the valence shell electron configuration of each O atom is $2s^2$ $2px^2$

$2py^1 2p_z^1$. It is conventional to take z axis as the internuclear axis or molecular axis. Along the molecular axis, overlap of $2P_z$ orbital of two O atoms occur with cylindrical symmetry thus forming a 1 ERQG

The remaining two $2p_y$ orbitals of two O atoms cannot overlap to the IXOO H[WHQW OLNH D 1 ERQG DV WKH\ GR QRW KDYH F\OLQGULFDO V\PPHWU\ DURXQG the internuclear axis. Instead, $2p_y$, orbitals overlap laterally (sideways) above and below the axis and share the pair of electrons. The bond formed by lateral overlap of p orbitals above and below the axis together is called a Œ 3L ERQG. Since $2p_y$ orbitals are perpendicular to $2p_z$ RUELWDOV Œ ERQG formed is perpendicular to the σ bond. Thus bonding in oxygen molecule is represented as in fig. 1.8(a).

There are two bonds in O2 PROHFXOH 2QH RI ZKLFK LV D 1 ERQG DQG DQRWKHU LV Œ ERQG

Similarly, in N_2 molecule, 3 bonds are present between 2N atoms. The nature of orbital overlaps in the 3 bonds can be considered as in fig. 1.8(b).

WR WKH 1 ERQG 7KXV ERQGLQJ LQ R[\JHQ PROHFXOH LV UHSUHVHQWHG DV LQ)LJ

1.8(a).

There are two bonds in O_2 PROHFXOH 2QH RI ZKLFK LV D 1 ERQG DQG DQRWKHU LV Œ ERQG Similarly, in N_2 molecule, 3 bonds are present between 2N atoms. The nature of orbital overlaps in the 3 bonds can be considered as in Fig. 1.10.

The valence electronic configuration of nitrogen atom is $2s^2$ $2p_x^1$ $2p_y^1$ $2p_z^1$.

Cylindrically symmetrical overlap of two $2p_z$ RUELWDOV JLYH D 1 ERQG DQG lateral overlap of two $2p_y$orELWDOV JLYH D Œ ERQG SHUSHQGLFXODU WR 1 ERQG Similarly, two $2p_z$ RUELWDOV ODWHUDOO\ RYHUODS WR JLYH DQRWKHU Œ ERQG ZKLFK LV SHUSHQGLFXODU WR ERWK 1 Œ ERQGV

Based on the valence bond orbital overlap theory, the H_2O molecule is viewed to be formed by the overlap 1s orbital of a H atom with 2py orbitalRI 2 DWRP FRQWDLQLQJ RQH HOHFWURQ HDFK IRUPLQJ D 1 ERQG $QRWKHU 1 ERQG is also formed by the overlap of 1s orbital of another H atom with $2p_x$ orbital of O atom each containing an unpaired electron. The bond angle is therefore 90° (i.e : HOH bond angle is 90°), since $2p_x$ and $2p_y$ orbitals are mutually perpendicular to each other. However the actual bond angle value is found to be 104°. Therefore based on VB theory, pure orbital overlaps does not explain the geometry in H_2O molecule.

Similarly, in NH_3, according to VB theory each of N-H bonds are formed by the overlap of a 2p orbital of N and 1s orbital of H atoms respectively. Here again the bond angle of HNH bond is predicted as 90°. Since $2p_x$, $2p_y$ and $2p_z$ orbitals of N are mutually perpendicular. However the experimental bond angle of HNH bond value is found to be 107°.

1.6.1 Theory of Hybridization

The failures of VB theory based on pure orbital overlaps are explained agreeably based on the concept of hybridization of orbitals or mixing up of orbitals. There are three major processes that are considered to occur in hybridization of orbitals. These are:

i) Promotion of electrons to higher or similar energy levels

ii) Mixing up of various s,p,d,f orbitals to form the same number of new orbitals and

iii) Stabilization of the molecule through bond formations involving hybrid orbitals by release of certain amount of energy which compensates the energy requirement in the electron promotion process.

According to VB theory, Beryllium is expected to behave like a noble gas due to its filled shells, which in practice forms a number of compounds like BeF_2 and BeH_2 proving its bivalency. In case of Boron VB theory predicts univalency due to the presence of one unpaired electron but in practice Boron is trivalent since compounds as BCl_3, BH_3 etc. are found.

The stable state (Ground State) electronic configuration of C is ($2s^2 2p_x^1 2p_y^1$). Electronic configuration of C suggests only bivalency. But carbon forms over a million compounds in all of which carbon is tetravalent. This suggests only tetravalency. This deficiency is overcome by allowing for **promotion** (or) **the excitation** of an electron to an orbital of higher energy. Although for electron promotion energy is needed, if that energy is recovered back during a covalent bond formation, or by a bond with a greater strength or by many number of bonds formation, then the electron promotion becomes energetically allowed and assumed to take place initially. In carbon, promotion of an electron to an orbital which is close to itself with an empty orbital of only slightly higher energy which is the $2p_z$ orbital can take place. Then the electron pair is unpaired itself by absorbing the required energy available by the atom from its surrounding and one of the electrons in the original orbital 2s or 2p shifts to the empty higher energy orbital.

Thus promotion of an electron leads to four unpaired electron in the excited state electronic configuration of carbon atom. Each electron can now be utilised to form a covalent bond by sharing an electron coming IURP WKH FRPELQLQJ DWRP 7KXV IRXU 1 FRYDOHQW ERQGV DUH SRVVLEOH HDFK with equivalent strength and overlapping tendency. Further, chemical and physical evidences reveal the four bonds of carbon to be equivalent and that they are tetrahedrally oriented. The promotion of an electron from 2s to 2p orbital leads to four half-filled orbitals which can form four bonds leading to greater energy lowering. This energy is more than the initial energy required for the promotion of 2s electron to 2p orbital.

Hybridization (mixing of orbitals)

After an electron promotion the 4 electrons are not equivalent, since one of them involves with an s orbital while the other three involve with p orbitals. To explain the equivalence of the four bonds, the concept of hybridization is introduced.

Dissimilar orbitals like s,p,d with one or many numbers, with nearly the same energy on the same atom may combine or mix completely to form an equal number of equivalent energy new orbitals with properties of their own. This is called **as hybridization** of orbitals. The new orbitals formed are known as hybrid orbitals and these orbitals possess the properties of the pure orbitals that are mixed to form them. The hybrid orbitals of an atom are symmetrically distributed around it in space. Essentially, mixing up of orbitals to form new orbitals explains the different geometries of many compounds like CH_4, SF_6 etc.

1.7 Concept of Resonance

According to the concept of resonance whenever a single Lewis structure cannot describe a molecular structure accurately, a number of structures with similar energy, positions of nuclei, bonding and non-bonding pairs of electrons are considered to represent the structure. Each such structure is called as canonical structure. A resonance hybrid consists of many canonical structures. All the canonical structures are equally possible to represent the structure of the molecule.

For example, in ozone (O_3) molecule, the two canonical structures as shown below and their hybrid represents the structure of O_3 more accurately. Resonance is represented by a double headed arrow placed between the canonical structures. There are two canonical forms of O_3.

The resonance structures are possible for molecular ions also. For example, consider resonance in CO_3^{2-} ion:-

The single Lewis structure based on the presence of two single bonds and one double bond between each carbon and oxygen atoms is inadequate to represent the molecule accurately as it represents unequal bonds. According to experimental findings all carbon to oxygen bonds in CO_3^{2-} are equivalent. Therefore the carbonate ion is best described as a resonance hybrid of the canonical forms as shown in Fig. 1.9 b.

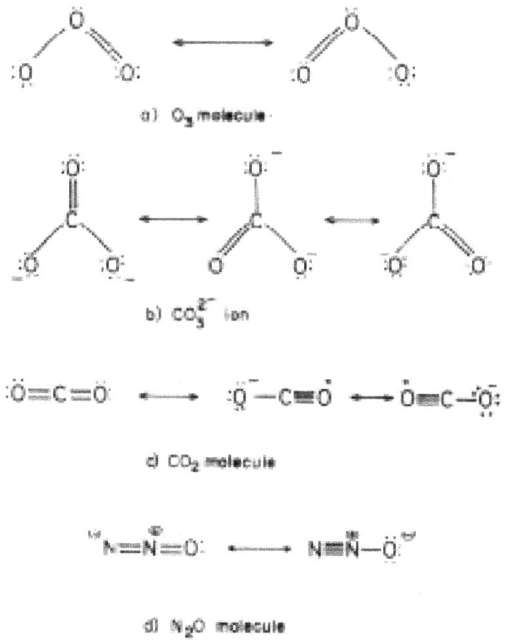

a) O_3 molecule

b) CO_3^{2-} ion

c) CO_2 molecule

d) N_2O molecule

Fig. 1.9 Resonance structures of (a) Ozone (b) Carbonate ion(c) Carbon dioxide (d) Nitrousoxide

There are three canonical forms of CO_3^{2-}.

Structure of CO_2 molecule is also an example of resonance, the experimental C-O bond length is found to be shorter than C-O single bond length and longer than C=O bond length and lies intermediate in value between a pure single and a pure double bond lengths. Also the two C=O bond length in the CO_2 molecule are equivalent and the properties of the two bonds are also the same. Therefore, a single lewis structure cannot depict the structure of CO_2 as a whole and it is best described as a resonance hybrid of the canonical forms given in Fig. 10.9c.

In N_2O molecule which is a linear molecule, structures with charges on atoms can be written similar to CO_2.

- Here also the experimental bond length of N-N-bond lies between a double and triple bond and that of N-O bond length lies between a single and a double bond. Therefore N_2O exists as a hybrid structure of the two canonical forms with a linear geometry.

1.8 Co-ordinate-covalent bonding or Dative bonding

The electron contributions of combining atoms in a covalent bond are generally equal. In each shared pair of electrons one electron is contributed from each atom of the bond. However in some bond formation, the whole of the shared pair of electrons comes from only one of the combining atoms of the bond, which is to referred as the donor atom. The other atom which does not contribute the electron to the shared pair but tries to pull the pair of electron towards itself is called as the acceptor atom. The bond thus formed is between the donor and acceptor atoms is called as the **co-ordinate or co-ordinate - covalent or dative bond.**

A coordinate bond is showed as an arrow which points from the donor to the acceptor atom. In some cases, the donated pair of electron comes from a molecule as a whole which is already formed to an already formed acceptor molecule as a whole.

For Example, coordination bond between H_3N: and BF_3 molecules. The molecule, ammonia (donor) which gives a pair of electron (lone pair) to BF_3 molecule which is electron deficient (acceptor) which has an empty orbital to accommodate the pair of electrons. Thus a dative bond is formed and the molecule as a whole is represented as H_3N: BF_3 (Fig. 1.10a).

When Proton is added to ammonia, a pair of electron is donated by nitrogen to proton and then proton shares the electron pair to form coordinate covalent bond.

Similarly in (NH_4Cl) ammonium chloride, covalent - coordinate bond exists in NH^{4+} ion only and Cl^- ion exists as it is.

Few examples of covalent - coordinate bond:

In nitro methane (CH_3-NO_2), one of the N-O-bond exists in a covalent coordinate type.

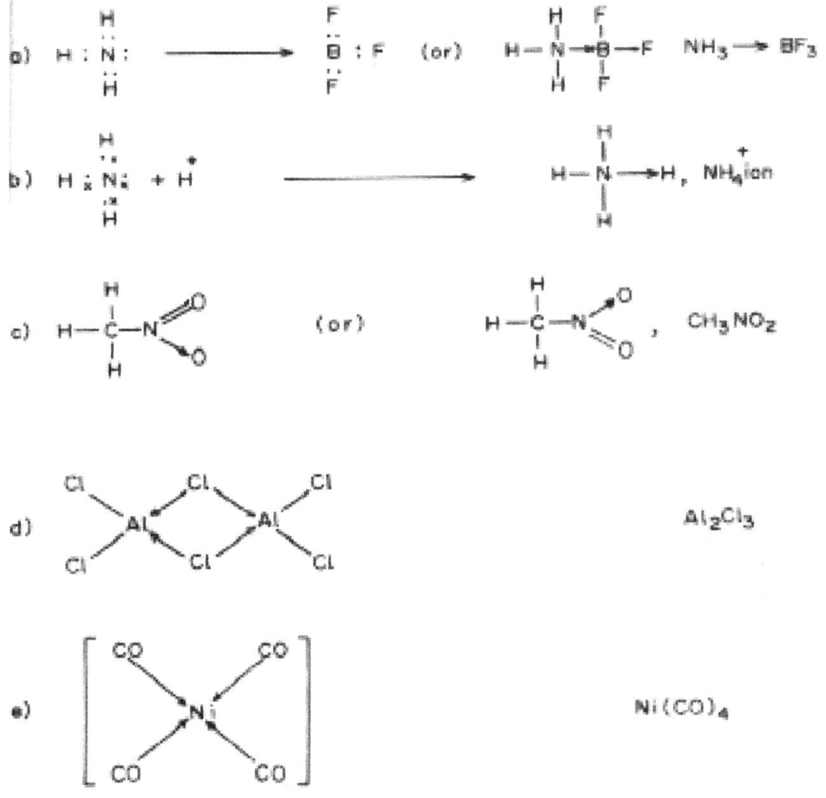
a)
(or)
$NH_3 \rightarrow BF_3$
b)
NH_4^+ ion
c)
(or)
CH_3NO_2
d)
Al_2Cl_3
e)
$Ni(CO)_4$

Fig. 1.10 Coordinate bonding in

(a) ammonia-borontrifluoride (b) ammonium ion (c) nitromethane (d) Aluminium chloride and (e) Nickel tetracarbonyl

Aluminum chloride Al_2Cl_6 (dimeric form)

Lone pairs of electron from chlorine are donated to electron deficient aluminum atoms in such a way that dimers of $AlCl_3$ are formed easily (Fig. 1.10d). The two chlorine atoms act as bridge to link the two Aluminum atoms.

In some complex ion formations, if the central transition metal-ion has empty `d' orbitals then lone pair of electrons from neutral molecules or anions are donated resulting in the formation of coordination bonds. Example: In Nickel tetra carbonyl, the four bonds between central Ni atom and the carbonyl ligands are mainly covalent -coordinate type. This complex exists in square planar geometry.

SUMMARY

- Chemical bonding is defined and Kossel-Lewis approach to understand chemical bonding by using the octet rule is studied. Except helium, atoms share or transfer valence electrons to attain the stable octet shell as the electronic configuration.
- Ionic bonding results due to complete electron transfer from electropositive elements to electronegative elements forming cation and anion. Electrostatic force of attraction between ions describes the ionic bonding. Mutual sharing of electrons between the two atoms result in covalent bonding. The directional characters, partial

ionic character by the pure orbital overlaps are also studied with suitable examples.

- The geometry of simple molecules are predicted using the postulates of VSEPR model $BeCl_2$: linear; CH_4 : Tetrahedral; BCl_3 : trigonal; PCl_5 : trigonal-bipyramidal; SF_6 : Octahedral.
- The concept of hybridization of C, N, O are learnt. σ and π bonds are studied and differentiated. Resonance in benzene, carbonate ion, molecules are understood.

Formation of coordinate covalent (dative) bonding between Lewis acids and electron donors are studied. Al_2Cl_6 is covalent but in water, it is ionic. Coordinate-covalent bonding in $Ni(CO)_4$ is also understood.

CHAPTER – 2

COLLIGATIVE PROPERTIES

OBJECTIVES

After Studying this Chapter you will able to:

- *To know about colligative properties and the scopes to determine molar mass of the non-volatile solute.*
- *To define Raoult's law and relate the relative lowering of vapour pressure to the molar mass of the solute in the solution.*
- *To determine experimentally the depression in freezing point by Beckmann method and use it to find the molar mass of a non-volatile solute.*
- *To know cottrell's method of elevation of boiling point and use it know the molar mass of a nonvolatile solute.*
- *To understand the concept of osmosis and to find the molar mass of a solute using osmotic pressure.*
- *To explain abnormal colligative properties as due to association and dissociation of solute molecules.*

2.1 Colligative Properties and its Scope

A solution may be considered as a homogeneous (single phase) mixture of two or more substances. It is said to be `binary' if two substances are present and `ternary' if three substances are present and `quaternary' if four substances are being present etc. In a binary solution, the component present in larger amount is called as solvent and the component in smaller amounts is called as solute. Solvent and solute together make a solution. In dilute solutions, very small amount of the solute is present.

A colligative property of a solution depends purely on the number of particles dissolved in it, rather than on the chemical nature of the particles. The colligative properties can be regarded as the properties of the solvent in a given solution. Generally, the solute is considered as non-volatile. The various colligative properties are as below:

i. /RZHULQJ RI YDSRXU SUHVVXUH RI WKH VROYHQW ûS

ii. (OHYDWLRQ RI ERLOLQJ SRLQW RI WKH VROYHQW û7$_b$)

iii. 'HSUHVVLRQ RI IUHH]LQJ SRLQW RI WKH VROYHQW û7$_f$)

iv. OsPRWLF SUHVVXUH Œ

The important scope of the measurement of colligative properties lies on its use to determine the molar mass of the non-volatile solute dissolved in the dilute solution.

/RZHULQJ RI 9DSRXU 3UHVVXUH ûS

If we take a pure liquid in a closed container, we find that a part of the liquid evaporates and fills the available space with its vapour. The vapour exerts a pressure on the walls of the container and exists in equilibrium with the liquid. This pressure is referred as the vapour pressure of the liquid.

When a non-volatile solute is dissolved in the solvent so that a dilute and homogeneous solution results, then again the vapour pressure of the solution will be made up of entirely from the solvent since the solute does not evaporate. This vapour pressure of the dilute solution is found to be lower than the vapour pressure of the pure solvent.

From Fig.2.1 it may be seen the surface of a dilute solution is partly occupied by solute molecules, thereby the number of solvent molecules at the surface being reduced. Consequently the vapour pressure of the solvent molecules gets lowered on the surface of the solution.

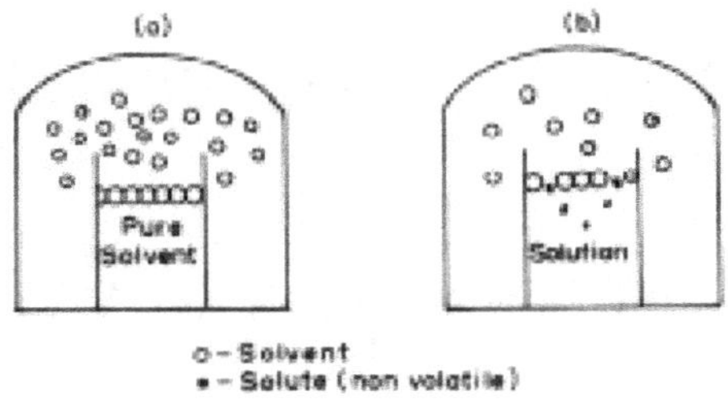

Fig. 2.1 Effect of solute in the solution on the vapour pressure

2.3 Raoult's Law

The relationship between the vapour pressure of the solution and its concentration is given by a French chemist named Francois Marie Raoult (1886). According to Raoult's law, at constant temperature the vapour pressure of the solution (p) is directly proportional to the molefraction of the solvent (X_1) present in the solution. That is, p .;1 (or) $p = kX_1$ where k is the proportionality constant. The value of k is known as follows: For a pure solvent, X_1=1.0 and p becomes p° corresponding to the vapour pressure of the pure solvent. Thus, p° = k (1.0). Substituting the value of k,

$$p = p^\circ X_1 \qquad \dots 2.1$$

Equation 2.1 is generally known as Raoult's law.

When n_1 and n_2 are the number of moles of solvent and solute present in the solution, the molefraction of the solvent $X_1 = n_1/(n_1 + n_2)$ and the mole fraction of solute $X_2 = n_2/(n_1+n_2)$. Also, $X_1 + X_2 = 1.0$

If W_1 and W_2 are the weights of solvent and solute present, then $n_1 = W_1/M_1$ and $n_2 = W_2/M_2$. M_1 and M_2 are the molar masses of solvent and solute respectively.

It is generally observed that p is lower than P°. The lowering of vapour pressure of the solvent in the solution equals to (p° - S ûS

The relative lowering of the vapourpressure is defined as the ratio of the lowering of vapour pressure to the vapour pressure of the pure solvent. Thus relative lowering of vapour pressure is given by

$$\frac{p^\circ - p}{p^\circ} = \frac{ûS}{p^\circ}$$

$$\frac{p^\circ - p}{p^\circ} = \frac{p^\circ - p^\circ X_1}{p^\circ} \qquad \text{since } p = p^\circ X_1$$

$$\frac{P^\circ(1-X_1)}{p^\circ} = 1-X_1 = X_2 \dots \text{since } \{X_1+X_2 = 1\}$$

$$\therefore \frac{p^\circ - p}{p^\circ} = X_2 \qquad \dots 2.2$$

Equation 2.2 represents the mathematical from of Raoult'slaw. Thus, the statement of Raoult's law for dilute solutions containing non-volatile non-electrolyte solute is: Relative lowering of vapour pressure is equal to the mole fraction of the solute. Since mole fraction of the solute (X_2) is given by $n_2/(n_1+n_2)$, the quantity $(p°-p)/p°$ depends upon the number of moles or molecules of the solute in solution and not on its chemical nature. Thus, relative lowering of vapour pressure is a colligative property.

2.3.1 Determination of molecular weights from relative lowering of vapour pressure

In dilute solutions, the number of moles of solvent (n_1) is large compared to the number of moles of solute (n_2) and thus ($n_1 + n_2$) can be approximated to n_1 and x_2 becomes equal to n_2/n_1. Measurement of lowering of vapour pressure, M_2 the molar mass of the solute can be determined using equation 2.3.

$$\text{Thus} \quad \frac{\hat{U}p}{p°} = \frac{n_2}{n_1} = \frac{W_2.M_1}{M_2.W_1} \qquad \ldots 2.3$$

Substituting for n_1 and n_2 as W_1/M_1 and W_2/M_2 we get $\hat{u}S\ S° = M_1.W_2/W_1.M_2$. Knowing M_1, W_1 and W_2 And from the

2.3.2 Experimental determination of relative lowering of vapour pressure Dynamic method (or) Ostwald-Walker method

This method is based on the principle that when dry air is successively passed through a series of containers possessing solution and pure solvent respectively, the air becomes saturated with the solvent vapours and an equal amount of weight loss in solution and solvent containers takes place.

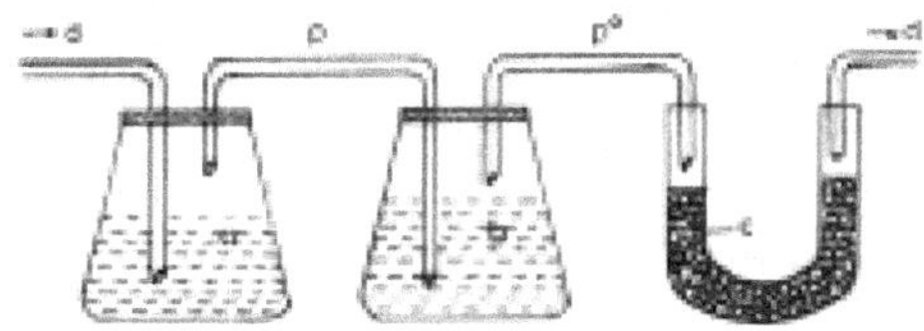

Fig. 2.2 Ostwald - walker apparatus

In Fig. 2.2 the first chamber (a) contains a weighed amount of the solution under examination and the next chamber (b) contains a weighed amount of the pure solvent. A weighed amount of anhydrous and dry calcium chloride is taken in the U-tube (c) connected at the end. The chambers and the U-tube are connected by a series of delivery tubes(d) through which air is passed. The dry air is first allowed to pass through the solution chamber until the air is saturated with the solvent vapour to maintain the vapour pressure of the solution `p'. Consequently, a loss in weight of the solution results in the solution chamber since some amount solvent molecules have evaporated. When this air is allowed to pass through the pure solvent chamber some more solvent vapour gets in stream with air, until the vapour pressure of pure solvent p°, is maintained. This happens so because p° is greater than p. Consequently, the weight loss registered in the solvent chamber is proportional to the (p°-p) quantity.

The weight loss in solution chamber . S

The weight loss in solvent chamber . S°-p

Sum of the loss in weights of solution

and solvent chamber . (p+p°-p) . S°

When the air saturated with solvent vapours is passed through

$CaCl_2$ U-tube, the solvent vapours are absorbed and the dry air gets out. The gain in weight of the $CaCl_2$ U-tube should be equal to the total loss in weight of solution and solvent chambers, which is inturn proportional to p°.

$$\frac{\text{Loss in weight of the solvent chamber}}{\text{Gain in weight of } CaCl_2 \text{ tube}} = \frac{p^\circ - p}{p^\circ}$$

= relative lowering of the vapour pressure

Thus, using the experimental $(p^\circ-p)/p^\circ$ values and applying Raoult's law, the molecular weight of the solute can be determined.

2.4 Depression of freezing point of dilute solution

Freezing point is the temperature at which solid and liquid states of a substance have the same vapour pressure. According to Raoult's law, addition of a non-volatile solute to solvent lowers the vapour pressure of the solvent and hence, the vapour pressure of pure solvent is greater than the vapour pressure of solution. Thus the temperature at which the solution and its solid form existing in equilibrium and possessing the equal vapour pressures, is lowered. That is, the freezing point of solution is lowered. The lowering of the freezing point of the solution from that of the freezing point of the pure solvent is known as depression in freezing point of the solution.

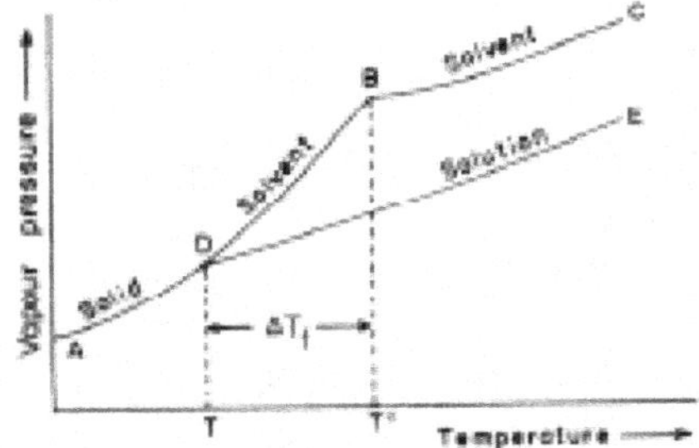

Fig. 2.3 Vapour pressure - temperature curves for depression in freezing

point

Consider the vapour pressure curves shown in Fig.2.3. Generally when the temperature of a solid substance that is used as the solvent is raised, the vapour pressure also raises. AB curve depicts this. Similarly curve BC represents the increase in vapour pressure of the liquid solvent with increase in temperature. Curves AB and BC meet at B corresponding to T^o temperature which is the freezing point of the pure solvent. At T^o, the vapour pressure of the liquid and solid states of the solvent are equal at B. Since the vapour pressure of the solution is always lower than that of its pure solvent, the vapour pressure curve of the solution DE always lie below that of the pure solvent.

D is the point of intersection of the vapour pressure curves of solution and pure solvent. The temperature at D is the freezing point of the solution and is seen to be lower than T^o. The depression in freezing point is û7f = T^o7 7KH PHDVXUHG GHSUHVVLRQ LQ IUHH]LQJ SRLQW û7_f) is found to be directly proportional to the molality (m) of the solute in solution. That is, û7f . P RU û7f = Kf m, where K_f is called as the **cryoscopicconstant(or) molal freezing point depression constant.** `K_f' is defined as thedepression in freezing point produced when one mole of solute is dissolved in 1 kg solvent. It is also the depression in freezing point of one molal solution. Freezing point depression of a dilute solution is found to be directly proportional to the number of moles (and hence the no.of molecules) of the solute dissolved in a given amount of the solvent.

Also û7f is independent of the nature of the solute as long as it is non-volatile. Hence depression in freezing point is considered as a colligative property.

Determination of molecular weight from depression in freezing point

$$\text{û7f} = K_f . m$$

Where m = molality

$$M = \frac{n_2}{W_1} \text{ and } n_2 = \frac{W_2}{M_2}$$

W_1= Weight of the solvent in Kg;

M_2= Molecular weight of solute

$$\therefore m = \frac{W_2}{M_2 W_1}$$

$$\therefore \text{û7}_f = K_f \frac{W_2}{M_2 . W_1}$$

$$\therefore M_2 = \frac{K_f . W_2}{\text{û7}_f . W_1} \quad \frac{K.kg\ mol^{-1} g}{K.kg}$$

$$M_2 = \frac{K_f}{\text{û7}_f} . \frac{W_2}{W_1} \ g\ mol^{-1}$$

Thus the molecular weight of the solute can be calculated.

2.4.1 Beckmann Method

Beckmann thermometer is used to measure small temperature changes in the freezing point of pure solvent and solution. Beckmann thermometer is not used in determining the absolute value of freezing temperature of the solvent or that of the solution. It is therefore called a differential thermometer. Temperature differences of even 0.01K can easily be measured.

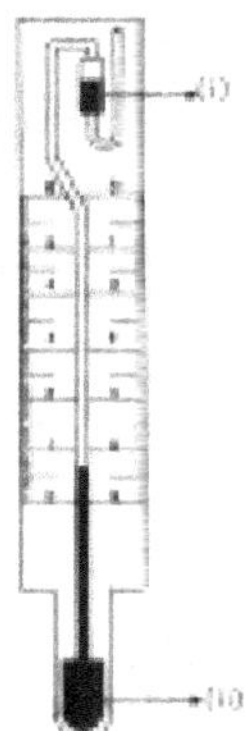

Fig. 2.4 Beckmann thermometer

Beckmann thermometer (Fig.2.4) consists of a large thermometer bulb at the bottom of a free capillary tube (ii) which is connected to a reservoir of mercury (i) placed at the top. As the capillary has fine bore, a small change of temperature causes a considerable change in the height of mercury column (level) in the capillary. The whole scale of a Beckmann thermometer covers only about 6K. Initially the level of mercury in the capillary should be on the scale. This is achieved by transferring mercury from the lower bulb to the reservoir and viceversa. When the Beckmann thermometer is used at high temperatures, some of the mercury from the thermometer bulb is transferred into the upper reservoir. At lower temperature mercury from the reservoir falls down in to the thermometer bulb.

Measurement of freezing point depression by Beckmann method

A simple Beckmann apparatus is shown in Fig.2.5. It consists of a freezing tube (a) with a side arm (c) through which a known amount of a solute can be introduced. A stopper carrying a Beckmann thermometer (b) and a stirrer (d) is fitted in to the freezing tube. To prevent rapid cooling of the contents of the freezing tube, A, a guard tube (e) surrounds the tube so that there is an air space between a and e. This assembly, as a whole, is placed in a wide vessel V which contains a freezing mixture (f) maintaining a low temperature around 5°C below the freezing point of the pure solvent.

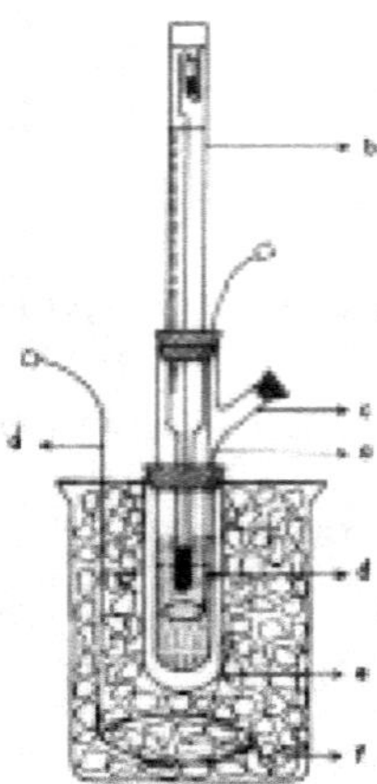

Fig. 2.5 Apparatus for Beckmann method

A known weight of the pure solvent is placed in the tube (a). It is cooled with gentle and continuous stirring. As a result of super cooling, the temperature of the solvent will fall by about 0.5°C below its freezing point. Vigorous stirring is then set in when solid starts separating and the temperature rises to the exact freezing point. This temperature remains constant, for some time, until all the liquid solvent gets solidified and is noted as T^o.

The tube (a) is taken out, warmed to melt the solid and a known weight of the solute is added through the side arm (c). When the solute is dissolved in to the solvent forming a solution, the tube (a) is put back in to the original position and the freezing point of the solution (T) is redetermined in the same manner as before. The difference between the two readings gives the freezing point depression (û7f).

'HSUHVVLRQ LQ IUHH]LQJ SRLQW û7f = T°-T. From this value, the molecular mass of the non-volatile solute can be determined using the expression and known K_f value.

$$M_2 = \frac{Kf}{\hat{u}7_f} \cdot \frac{W_2}{W_1} \quad \text{... 2.4}$$

Table 2.1 Molal Depression (cryoscopic) constants, K_f (One mole of solute per 1000 grams of solvent)

Solvent	F. Pt. K	K_f (K.kg.mole^{-1})
Acetic acid	289.60	3.90
Bromoform	281.30	14.30
Benzene	278.53	5.10
Cyclohexane	279.55	20.20
Camphor	451.40	37.70
Naphthalene	353.25	7.00
Nitrobenzene	278.70	6.90
Phenol	314.10	7.27
Water	273.00	1.86

2.5 Elevation of boiling point of dilute solutions

The boiling point of a pure liquid is the temperature at which its vapour pressure becomes equal to the atmospheric pressure. Since the vapour pressure of a solution is always lower than that of the pure solvent, it follows that the boiling point of a solution will always be higher than of the pure solvent.

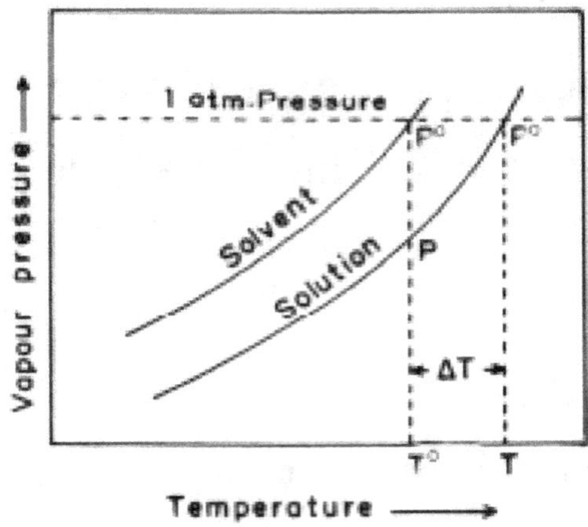

Fig. 2.6

In the Fig.2.6, the upper curve represents the vapour pressure - temperature dependance of the pure solvent. The lower curve represents the vapour pressure - temperature dependance of a dilute solution with known concentration. It is evident that the vapour pressure of the solution is lower than that of the pure solvent at every temperature. The temperature T^o gives the boiling point of the pure solvent and T the boiling point of the pure solution. This is because at these temperatures (T^o, T) the vapour pressures of pure solvent and solution becomes equal to the atmospheric pressure.

7KH HOHYDWLRQ RI ERLOLQJ SRLQW û$7_b = T - T^o$.

Elevation of boiling point is found directly proportional to the molality of the solution (or) inturn the number of molecules of solute. Also it is independent of the nature of the solute for a non-volatile solute. Hence, boiling point elevation is a colligative property.

Thus it may be written as

û7b . P	... 2.5

Determination of molecular weight from boiling point elevation

By measuring the boiling point elevation of a solution of a known concentration, it is possible to calculate molecular weight of a non-volatile non-electrolyte solute.

û7b . P

∴ û7$_b$= $K_b m$ ... 2.6

The proportionality constant K_b is characteristic of the solvent and it is called the **molal boiling point elevation constant** or **ebullioscopicconstant**. It is defined as the elevation of boiling point of one molalsolution.

When n_2 moles of the solute is dissolved in W_1 kg of the solvent, the molality is given by n_2/W_1.

Since W_2, is the weight of the solute, we can calculate the molecular weight of the solution using the following expression.

$$\therefore M_2 = \frac{K_b \cdot W_2}{\text{û7b} \cdot W_1}$$

Table 2.2 Molal Elevation (Ebullioscopic) constants (One mole of solute per 1000 grams of solvent)

Solvent	B. Pt K	K_b (K.kg.mole^{-1})
Water	373.00	0.52
Benzene	353.10	2.57
Methanol	337.51	0.81
Ethanol	351.33	1.20
Carbon tetra chloride	349.72	5.01
Chloroform	334.20	3.88
Acetic acid	391.50	3.07
Acetone	329.15	1.72
Carbon disulphide	319.25	2.41

Phenol	455.10	3.56

2.5.1 Determination of elevation of boiling point by Cottrell's Method

The apparatus (Fig.2.7) consists of a boiling tube (a) which is graduated and contains weighed amount of the liquid under examination. An inverted funnel tube (b) placed in the boiling tube collects the bubbles rising from a few fragments of a porous pot placed inside the liquid. When the liquid starts boiling, it pumps a stream of a liquid and vapour over the bulb of the Beckmann thermometer (f) held a little above the liquid surface. In this way, the bulb is covered with a thin layer of boiling liquid which is in equilibrium with the vapour. This ensures that the temperature reading is exactly that of the boiling liquid and that superheating is minimum. After determining the boiling point of the pure solvent, a weighed amount of the solute is added and procedure is repeated for another reading. The vapours of the boiling liquid is cooled in a condenser (C) which has circulation of water through (d) and (e). The cooled liquid drops into the liquid in (a).

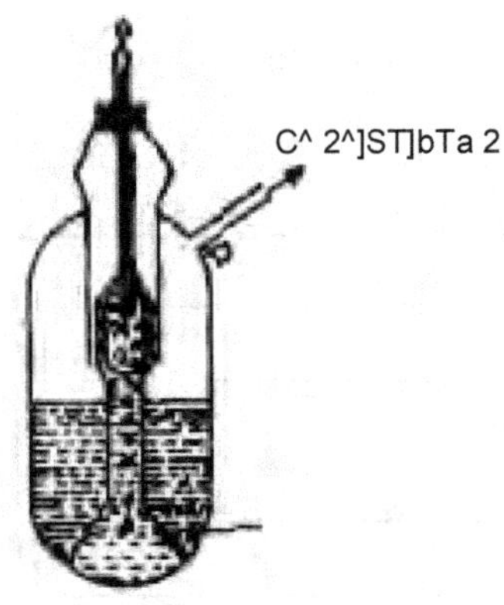

Fig, 2.7

2.6 Osmosis in solution

Spontaneous movement of solvent particles from a dilute solution or from a pure solvent towards the concentrated solution through a semipermeable membrane is known as **osmosis** (Greek word: 'Osmos' = to push).

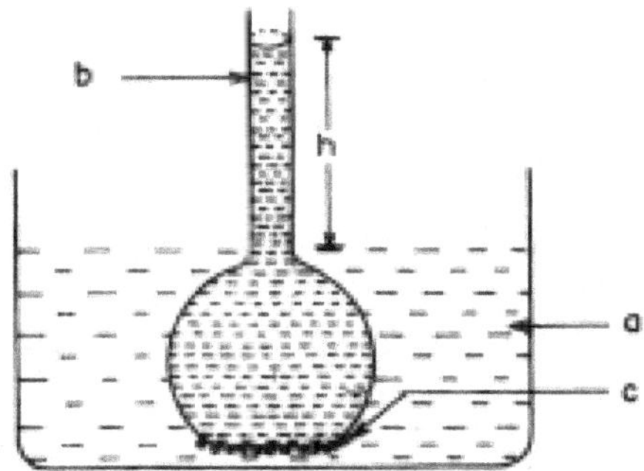

Fig. 2.8 Osmosis apparatus

Fig. 2.8 depicts Osmosis in a simple way. The flow of the solvent from its side (a) to solution side (b) separated by semipermeable membrane (c) can be stopped if some definite extra pressure is applied on the solution risen to height (h). This pressure that just stops the flow of solvent is called **osmotic pressure** of the solution. This pressure (π) has been found todepend on the concentration of the solution.

Osmosis is a process of prime importance in living organisms. The salt concentration in blood plasma due to different species is equivalent to 0.9% of aqueous sodium chloride. If blood cells are placed in pure water, water molecules rapidly move into the cell. The movement of water molecules into the cell dilutes the salt content. As a result of this transfer of water molecules the blood cells swell and burst. Hence, care is always taken to ensure that solutions that flow into the blood stream have the same osmotic pressure as that of the blood.

Sodium ion (Na^+) and potassium ions (K^+), are responsible for maintaining proper osmotic pressure balance inside and outside of the cells of organism. Osmosis is also critically involved in the functioning of kidneys.

&KDUDFWHULVWLFV RI 2VPRWLF 3UHVVXUH Œ

- It is the minimum external pressure which must be applied on solution side in order to prevent osmosis if separated by a solvent through a semi permeable membrane.
- A solution having lower or higher osmotic pressure than the other is said to be hypotonic or hypertonic respectively in respect to other solution.
- Two solutions of different substances having same osmotic pressure at same temperature are said to be isotonic to each other. They are known as isotonic solutions.

2.6.1 Osmotic pressure and concerned laws

Vant Hoff noted the striking resemblance between the behaviour of dilute solutions and gases. He concluded that, a substance in solution behaves exactly like gas and the osmotic pressure of a dilute solution is equal to the pressure which the solute would exert if it is a gas at the same temperature occupying the same volume as the solution. Thus it is proposed that solutions also obey laws similar to gas laws.

1. Boyle's - Vant Hoff law

7KH RVPRWLF SUHVVXUH Œ RI WKH VROXWLRQ DW

FRQVWDQW WHPSHUDWXUH LV directly proportional to the concentration (C) of the solution.

Œ .& DW FRQVWDQW 7

C = Molar concentration

2. Charle's - Vant Hoff law

At constant concentration the osmotic pressure UH Œ RI WKH VROXWLRQ LV directly proportional to the temperature (T).

Œ . 7 DW FRQVWDQW &&RPELQLQJ WKHVH WZR ODZV

Œ .& 7

(or) Œ &57 2.7

where R is the gas constant.

Determination of molecular weight by osmotic pressure measurement

The osmotic pressure is a colligative property as it depends, on the number of solute molecules and not on their identity.

Solution of known concentration is prepared by dissolving a known weight (W_2) of solute, in a known volume ($V\ dm^3$) of the solvent and its

RVPRWLF SUHVVXUH Œ LV PHDVXUHG DW URRP WHPSHUDWXUH 7

Since Œ = CRT

(or) $M_2 = \dfrac{W_2\,RT}{\text{Œ 9}}$ 2.8

Thus M_2, molecular weight of the solute can be calculated by measuring osmotic pressure value.

2.6.2 Determination of osmotic pressure by Berkley-Hartley method

The osmotic pressure of a solution can be conveniently measured by Berkley - Hartley method. The apparatus (Fig. 2.9) consists of two concentric tubes. The inner tube (a) is made of semipermeable membrane (c) with two side tubes. The outer tube (b) is made of gun metal which contains the solution. The solvent is taken in the inner tube. As a result of osmosis, there is fall of level in the capillary indicator (d) attached to the inner tube. The external pressure is applied by means of a piston (e) attached to the outer tube so that the level in the capillary indicator remains stationary at (d). This pressure is equal to the osmotic pressure (π) and the solvent flow from inner to outer tube is also stopped.

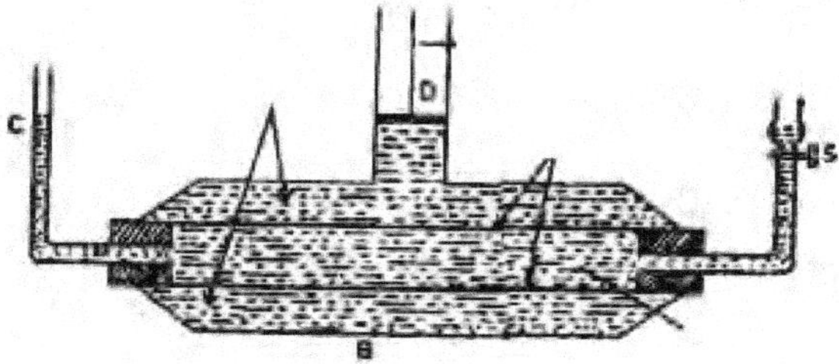

Fig. 2.9 Berkley - Hartley apparatus

Advantages of this Method

1. The osmotic pressure is recorded directly and the method is quick.
2. There is no change in the concentration of the solution during the measurement of osmotic pressure.
3. The osmotic pressure is balanced by the external pressure and there is minimum strain on the semipermeable membrane.

2.7 Abnormal Colligative Properties

The experimental values of colligative properties in most of the cases resemble closely to those obtained theoretically by their formula. However, in some cases experimental values of colligative properties differ widely from those obtained theoretically. Such experimental values are referred to as abnormal colligative properties.

The abnormal behaviour of colligative properties has been explained in terms of dissociation and association of solute molecules.

a. Dissociation of solute molecules

Such solutes which dissociate in solvent (water) i.e. electrolytes, show an increase in number of particles present in solution. This effect results in an increase in colligative properties obtained experimentally.

The Van't Hoff factor (i)

$$i = \frac{\text{Experimental colligative property}}{\text{Normal colligative property}}$$

I > 1 for dissociation. We can calculate the degree of dissociation (σ) using the equation.

$$\text{dissociation} = \frac{i - 1}{n - 1} \qquad \text{... 2.9}$$

where `n' is the total number of particles furnished by one molecule of the solute.

For example, sodium chloride in aqueous solution exists almost entirely as Na^+ and Cl^- ions. In such case, the number of effective particles increases and therefore observed colligative property is greater than normal colligative property.

b. Association of the solute molecules

Such solute which associate in a solvent show a decrease in number of particles present in solution. This effect results in a decrease in colligative properties obtained experimentally.

Here,

Experimental Colligative Property < Normal Colligative Property

∴ Vant Hoff factor

$$i = \frac{\text{Experimental Colligative Property}}{\text{Normal colligative property}}$$

i< 1 for association

Using this, the degree of association `. FDQ EH FDOFXODWHG IURP

$$\text{Association} = \frac{(1-i)n}{(n-1)} \quad \ldots 2.10$$

where `n' is the number of small molecules that associate into a single larger new molecule.

For example, molecules of acetic acid dimerise in benzene due to intermolecular hydrogen bonding. In this case, the number of particles is reduced to half its original value due to dimerization. In such case, the experimental colligative property is less than normal colligative property.

$$2\,(CH_3COOH) \rightleftharpoons (CH_3COOH)_2$$

Summary

Relationship between colligative properties and molecular mass of the non-volatile solute

1.	Relative lowering of vapour pressure $\frac{p^\circ - p}{p^\circ}$	The ratio of lowering of vapour pressure of the pure solvent	$\frac{P^\circ - P}{P^\circ} = \frac{W_2}{M_2} \cdot \frac{M_1}{W_1}$
2.	Elevation of ERLOLQJ SRLQW û7$_b$)	Boiling point of the solution is greater than the solvent	T-T$^\circ$ û7$_b$ û7$_b$= $\frac{W_2}{M_2W_1}$. Kb
3.	Depression in freezing point û7$_f$)	Freezing point of the solution is lower than solvent.	T$^\circ$-7 û7$_f$ û7$_f$= $\frac{W_2K_f}{M_2\,W_1}$
4.	Osmotic pressure Œ	Excess pressure applied on the concentrated solution side to stop the osmosis.	Œ &57
5.	Abnormal colligative property (i)	Due to dissociation and association of molecules, there is a change in the experimental colligative property value	Van't Hoff factor i = $\frac{\text{Observed colligative Property}}{\text{Theoretical colligative Property}}$

REFERENCES

1. Physical Chemistry by Lewis and Glasstone.

2. Physical Chemsitry by Maron and Prutton.

3. Physical Chemistry by P.L.Soni.

CHAPTER – 3

THERMODYNAMICS

OBJECTIVES

After Studying this Chapter you will able to:

- *To predict the possibility of a process.*
- *To differentiate system and surroundings from universe.*
- *To define various processes, properties; state and path functions; spontaneous and non-spontaneous;exo-and endo-thermic processes.*
- *To learn to interrelate work, heat and energy.*
- *To define Zeroth and first laws of thermodynamics.*
- *To measure changes in internal energy and enthalpy.*
- *To relate E and H.*
- *To determine enthalpy changes of various physical processes.*
- *To determine enthalpy changes in formation, combustion, neutralization.*
- *To understand non-conventional energy resources and to identify different renewable energy resources.*

3.1 Introduction

The term **thermodynamics** is derived from Greek word, **`Thermos'** meaning heat and **`dynamics'** meaning flow. Thermodynamics deals with the inter-relationship between heat and work. It is concerned with the interconversions of one kind of energy into another without actually creating or destroying the energy. **Energy** is understood to be the capacity to do work. It can exist in many forms like electrical, chemical, thermal, mechanical, gravitational etc. Transformations from one to another energy form and prediction of the feasibility (possibility) of the processes are the important aspects of thermodynamics.

As an illustration, from our common experience steam engines are seen to transform heat energy tomechanical energy, by burning of coal which is a fossil fuel. Actually, the engines use the energy stored in the fuel to perform mechanical work. In chemistry, many reactions are encountered that can be utilized to provide heat and work along with the required products. At present thermodynamics is widely used in physical, chemical and biological sciences focusing mainly on the aspect of predicting the possibility of the processes connected with each sciences. On the other hand, it fails to provide insight into two aspects: Firstly, the factor of time involved during the initial to final energy transformations and secondly, on the quantitative microscopic properties of matter like atoms and molecules.

3.2 Terminology used in Thermodynamics

It is useful to understand few terms that are used to define and explain the basic concepts and laws of thermodynamics.

System

Thermodynamically a system is defined as any portion of matter under consideration which is separated from the rest of the universe by real or imaginary boundaries.

Surroundings

Everything in the universe that is not the part of system and can interact with it is called as surroundings.

Boundary

Anything (fixed or moving) which separates the system from its surroundings is called **boundary**.

For example, if the reaction between A and B substances are studied, the mixture A and B, forms the system. All the rest, that includes beaker, its walls, air, room etc. form the surroundings. The boundaries may be considered as part of the system or surroundings depending upon convenience. The surroundings can affect the system by the exchange of matter or energy across the boundaries.

Types of systems

In thermodynamics different types of systems are considered, which depends on the different kinds of interactions between the system and surroundings.

Isolated system

A system which can exchange neither energy nor matter with its surroundings is called an isolated system. For example, a sample in a sealed thermos flask with walls made of insulating materials represents an isolated system (Fig.3.1).

Closed system

A system which permits the exchange of energy but not mass, across the boundary with its surroundings is called a closed system.

For example: A liquid in equilibrium with its vapours in a sealed tube represents a closed system since the sealed container may be heated or cooled to add or remove energy from its contents while no matter (liquid or vapour) can be added or removed.

Open system

A system is said to be open if it can exchange both energy and matter with its surroundings.

For eg.a open beaker containing an aqueous salt solution represents open system. Here, matter and heat can be added or removed simultaneously or separately from the system to its surroundings.

All living things (or systems) are open systems because they continuously exchange matter and energy with the surroundings.

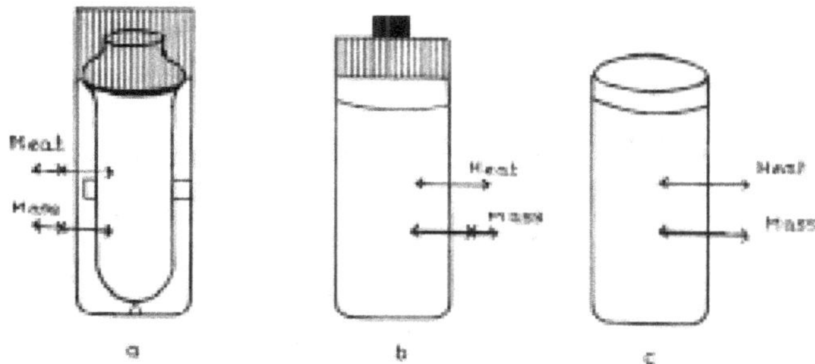

Fig. 3.1 Examples of

(a) isolated (thermos flask) (b) closed (closed beaker) (c) open (open beaker) systems

Homogeneous and Heterogeneous systems

A system is said to be **homogeneous** if the physical states of all its matter are uniform. For eg.mixture of gases, completely miscible mixture of liquids etc.

A system is said to be **heterogeneous**, if its contents does not possess the same physical state. For eg: immiscible liquids, solid in contact with an immiscible liquid, solid in contact with a gas, etc.

Macroscopic properties of system

The properties which are associated with bulk or macroscopic state of the system such as pressure, volume, temperature, concentration, density, viscosity, surface tension, refractive index, colour, etc. are called as macroscopic properties.

Types of macroscopic properties of system

Measurable properties of a system can be divided into two types.

Extensive properties

The properties that depend on the **mass** or **size** of the system are called as extensive properties. Examples: volume, number of moles, mass, energy, internal energy etc. The value of the extensive property is equal to the sum of extensive properties of smaller parts into which the system is divided. Suppose x_1 ml, x_2 ml,x_3 ml of 1,2,3 gases are mixed in a system, the total volume of the system equals to $(x_1 + x_2 + x_3)$ ml. Thus volume is an extensive property.

Intensive properties

The properties that are independent of the mass or size of the system are known as intensive properties. For eg.refractive index, surface tension, density, temperature, boiling point, freezing point, etc., of the system. These properties do not depend on the number of moles of the substance in the system.

If any extensive property is expressed per mole or per gram or per ml, it becomes an intensive property. For eg: mass, volume, heat capacity are extensive properties while density, specific volume, specific heat are intensive properties.

3.2.1 State functions

State of a system

A system is said to be in a particular physical state when specific values of the macroscopic properties of the system are known. For eg. The gaseous state of matter can be described by parameters like Pressure (P), Volume (V), Temperature (T) etc. The values of these parameters change when the matter is in liquid state. Thus, the **state of a system** is defined by specific measurable macroscopic properties of the system.

The **initial state** of system refers to the starting state of the system before any kind of interaction with its surroundings.

The **final state** of system refers to the state after the interaction of system with its surroundings. A system can interact with its surroundings by means of exchange of matter or heat or energy or all.

The variables like P,V,T, composition (no. of moles) `n' that are used to describe the state of a system are called as **state variables or statefunctions**. When the state of the system changes, the values of the statevariables of the system also change. Thus, state functions depend only on the initial and final states of system and not on how the changes occur. Also, if the values of state functions of a system are known, all other properties like mass, viscosity, density etc. of the system become specified.

For specifying a state of the system, it is not necessary to know all the state variables, since they are interdependent and only a few of them (state variables) are sufficient. A system which satisfies the conditions of thermal, mechanical and chemical equilibria and contains the macroscopic properties which are independent of time is said to be in **thermodynamicequilibrium**.

Thermodynamic equilibrium sets the condition that there should beno flow of heat from one portion or part of the system to another portion or part of the same system. i.e. temperature of the system remaining constant at every point of the system.

Mechanical equilibrium implies that there is no work done by oneportion or part of the system over another portion or part of the same system. i.e. Pressure of the system being constant at all its points.

Chemical equilibrium demands that the composition of one or morephases of chemicals present in the system should remain constant.

3.2.2 Thermodyanamic processes

A thermodynamic (physical or chemical) process may be defined as the pathway of series of intermediate changes that occur when a system is changed from initial to final state. Processes starting with the same initial state and ending at different final states correspond to different thermodynamic processes.

Different types of processes are commonly used in the study of thermodynamics.

Isothermal process is defined as one in which the temperature of thesystem remains constant during the change from its initial to final states. During the isothermal process, the system exchanges heat with its surroundings and the temperature of system remains constant.

Adiabatic process is defined as that one which does not exchange heatwith its surroundings during the change from initial to final states of the system.

A thermally and completely insulated system with its surroundings can have changes in temperature during transformation from initial to final states in an adiabatic process. This is because, the system cannot exchange heat with its surroundings.

Isobaric process is that process in which the pressure of the systemremains constant during its change from the initial to final state.

Isochoric process shows no change in volume of system during itschange from initial to final state of the process.

Cyclic process: The process which brings back the system to itsoriginal or initial state after a series of changes is called as cyclic process.

Spontaneous process are those that occur on their own accord. Forexample heat flowing from a hotter end of a metal rod to a colder end. In these processes, the transformation of the system from initial, to final state is favourable in a particular direction only. Many of the spontaneous processes are natural processes and are also, irreversible processes.

Non-spontaneous process are those that does not occur on their ownaccord. For example, although carbon burns in air evolving heat to form carbon dioxide, on its own carbon does not catch fire and an initial heat supply is required. Since many of the non-spontaneous processes are slow processes, they also exist as equilibrium processes.

Reversible process. In a reversible process the series of changescarried out on the system during its transformation from initial to final state may be possibly reversed in an exact manner.

This is possible when the changes are carried out very slowly in many smaller steps on the system during its change from initial to final state. By doing so, each of its intermediate state will be in equilibrium with its surroundings. Under such conditions the initial and final states of the system become reversible completely.

For example, when ice melts a certain amount of heat is absorbed. The water formed can be converted back to ice if the same amount of heat is removed from it. This indicates that many reversible processes are non-spontaneous processes also.

Irreversible Process

An irreversible process is one which cannot be retraced to the initial state without making a permanent change in the surroundings. Many of the spontaneous processes are irreversible in nature.

For eg. Biological ageing is an irreversible process. Water flowing down a hill on its own accord is an irreversible process.

Some of the characteristics of thermodynamically reversible and irreversible processes are compared as below:

Reversible process	Irreversible process
It is a slow process going through a series of smaller stages with each stage maintaining equilibrium between the system and surroundings.	In this process the system attains final state from the initial state with a measurable speed. During the transformation, there is no equilibrium maintained between the system and surroundings.
A reversible process can be made to proceed in forward or backward direction.	Irreversible process can take place in one direction only.
The driving force for the reversible process is small since the process proceeds in smaller steps.	There is a definite driving force required for the progress of the irreversible process.
Work done in a reversible process is greater than the corresponding work done in irreversible process.	Work done in a irreversible process is always lower than the same kind of work done in a reversible process.
A reversible process can be brought back to the initial state without making an change in the adjacent surroundings.	An irreversible process cannot be brought back to its initial state without making a change in the surroundings.

Exothermic and endothermic processes

When the thermodynamic process is a chemical reaction or a physical transformation, process is classified as either exothermic or endothermic depending on the nature of heat involved in the overall process. These two processes are differentiated as follows:

Endothermic process	Exothermic process
A process when transformed from initial to final states by absorption of heat is called as an endothermic process.	A process when transformed from initial to final states by evolution of heat is called as exothermic process.
The final state of the system possesses higher energy than the initial state. The excess energy needed is absorbed as heat by the system from the surroundings.	The final state of the system possesses lower energy than the initial state. The excess energy is evolved as heat. Example: All combustion processes are exothermic.
Generally in a physical transformation which is endothermic heat is supplied to bring about the initial to final state. Example: melting of a solid by supplying heat is an endothermic process.	If the physical transformation is exothermic heat is removed to bring about the initial to final state. Example: Freezing of a liquid at its freezing point is an exothermic process.

3.3 Nature of thermodynamic functions

The properties of a thermodynamic system depend on variables which are measurable and change in values when the state of the system changes. These variables are classified as state variables or state functions and path variables (or) path functions.

The **state functions** considered in a gaseous system like, P, V and T are called as state variables. A **state function** is a thermodynamic property of a system which has a specific value for each state of the system and does not depend on the path (or manner) in which a particular state is reached. Other than P,V,T there are other important thermodynamic properties existing as state functions like internal energy (U), enthalpy (H), free energy (G) etc. (The properties of U,H and G are to be studied later).

A **path function** is a thermodynamic property of the system whose value depends on the path or manner by which the system goes from its initial to final states. It also depends on the previous history of the system. For example, work (w) and heat (q) are some of the thermodynamic properties of the system that are path functions. Their values change when there is a change in manner in which the system goes from initial to final states.

3.4 Zeroth law of thermodynamics

Consider any two objects each maintained at different temperature, when brought in thermal contact with each other such that heat is exchanged until a thermal equilibrium is reached, then the two objects are considered to have equal temperatures. For example, if a beaker containing water and a thermometer are the two objects, while reading the temperature of the water in the beaker using the thermometer, a thermal equilibrium is reached between the two objects having a contact with each other. Also, when the temperatures of the thermometer bulb and that of water in the beaker are same, thermal equilibrium has said to be occurred.

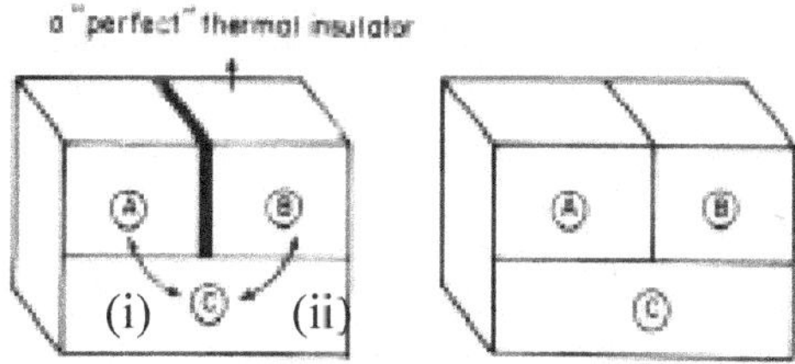

Fig. 3.2 i) A and B are in thermal equilibrium with C

ii)A,B and C are in thermal equilibrium with each other

Zeroth law of thermodynamics is also known as the law of thermal equilibrium. It provides a logical basis for the concept of temperature of a system. It can be stated as follows.

`If two systems at different temperatures are separately in thermal equilibrium with a third one, then they tend to be in thermal equilibrium with themselves'.

Conversely, the Zeroth law can be stated in another manner as,

`When two objects are in thermal equilibrium with the third object, then there is thermal equilibrium between the two objects itself'.

3.5 Work, heat and energy

In order to formulate the laws of

1 Torr = 1 mm of Hg

thermodynamics it becomes necessary to know the properties and nature of work (w) heat (q) and energy (u).

Work (w)

In thermodynamics work is generally defined as the force (F) multiplied by the distance of displacement(s). That is,

$$w = F.s.$$

Several aspects should be considered in the definition of work which are listed below:

(i) work appears only at the boundary of the system.

(ii) work appears during the change in the state of the system.

(iii) work brings in a permanent effect in the surroundings.

(iv) work is an algebraic quantity.

(v) work is a path function and it is not a state function.

Types of work

Many types of work are known. Some of the types of work are as follows:

(i) Gravitational work

This work is said to be done when a body is raised to a certain height against the gravitational field. If a body of mass `m' is raised through a height `h' against acceleration due to gravity `g', then the gravitational work carried out is `mgh'. In this expression, force is `mg' and the distance is `h'.

(ii) Electrical work

This type of work is said to be done when a charged body moves from one potential region to another. The electrical work is Q.V. if V is the potential difference causing the quantity of electricity 'Q' during its movement

(iii) Mechanical work

This type of work is associated with changes in volume of a system when an external pressure is applied or lowered. This pressure-volume work is also referred to as the mechanical work.

Heat

Like work, heat (q) is regarded in thermodynamics as energy in transit across the boundary separating a system from its surroundings. Heat changes result in temperature differences between system and surroundings. Heat cannot be converted into work completely without producing permanent change either in the system or in the surroundings. Some of the characteristics of heat (q) are:

(i) heat is an algebraic quantity.

(ii) heat is a path function and is not a state function.

(iii) heat changes are generally considered as temperature changes of the system.

Sign convention for heat (q) and work (w)

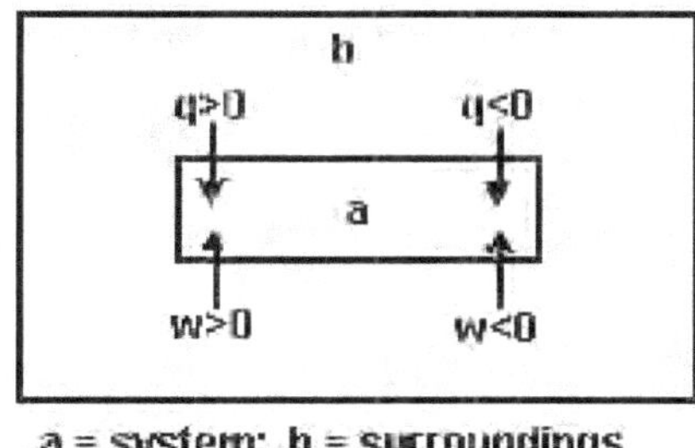

a = system; b = surroundings
q = heat; w = work

when,

(i) heat is absorbed by the system (or) heat is lost by

surroundings to the system: +q

(ii) heat is evolved by the system (or) heat is gained by surroundings: -q.

(iii) work is done by the system : -w

(iv) work is done on the system : +w

If heat (q) is supplied to the system, the energy of the system increases and `q' is written as a positive quantity. If work is done on the system, the energy of the system increases and `w' is written as a positive quantity. When w or q is positive, it means that energy has been supplied to the system as work or as heat. In such cases internal energy (U) of the system increases. When w or q is negative, it means that energy is lost by the system as work or as heat. In such cases, the internal energy (U) of the system decreases.

Energy `U'

Energy is easily, defined as the capacity to do work. Whenever there is a change in the state of matter of a system, then there is a change in energy û8 of the system. For example energy changes are involved in processes like melting, fusion, sublimation, vaporization etc. of the matter in a system. Energy (U) exists in many forms. Kinetic energy (K.E.) arises due to motion of a body and potential energy (P.E.) arises due to its position in space.

In chemical systems, there are two types of energy available. The energies acquired by the system like electrical, magnetic, gravitational etc. and termed as external energies of the system. The internal energy is generally referred to as the energy (U) of a thermodynamic system which is considered to be made up of mainly by P.E. and K.E.

Characteristics of energy (U) are:

(i) U is a state function. Its value depend on the initial and final states of the system.

(ii) U is an extensive property. Its magnitude depend on the quantity of material in the system.

(iii) U is not a path function. Its value remains constant for fixed initial and final states and does not vary even though the initial and final states are connected by different paths. In S.I. system the unit of energy is Joules `J' or kJ.

3.6 First law of thermodynamics

First law of thermodynamics is also known as the law of conservation of energy which may be stated as follows:

"Energy may be converted from one form to another, but cannot be created or be destroyed".

There are many ways of enunciating the first law of thermodynamics. Some of the selected statements are given below:

(i) "Energy of an isolated system must remain constant although it may be changed from one form to another".

(ii) "The change in the internal energy of a closed system is equal to the energy that passes through its boundary as heat or work".

(iii) "Heat and work are equivalent ways of changing a system's internal energy".

(iv) "Whenever other forms of energies are converted into heat or vice versa there is a fixed ratio between the quantities of energy and heat thus converted".

Significance of first law of thermodynamics is that, the law ascertains an exact relation between heat and work. It establishes that ascertain quantity of heat will produce a definite amount of work or vice versa. Also, when a system apparently shows no mechanical energy but still capable of doing work, it is said to possess internal energy or intrinsic energy.

3.7 Enthalpy

In chemistry most of the chemical reactions are carried out at constant pressure. To measure heat changes of system at constant pressure, it is useful to define a new thermodynamic state function called Enthalpy `H'.

H is defined as sum of the internal energy `U' of a system and the product of Pressure and Volume of the system.

That is,

$$H = U + PV$$

Characteristics of H

Enthalpy, H depends on three state functions U, P, V and hence it is also a state function. H is independent of the path by which it is reached. Enthalpy is also known by the term `heat content'.

3.7.1 Relation between enthalpy `H' and internal energy `U'

When the system at constant pressure undergoes changes from an initial state with H_1, U_1, V_1, P parameters to a final state with H_2, U_2,

V_2, P

û+ = $(H_2 - H_1) = (U_2 - U_1) + P(V_2 - V_1)$

i.e. û+ û8 3û9

&RQVLGHULQJ û8 T - w or q - 3û9 DVVXPLQJ S - 9 ZRUN û8 3û9 becomes equal to `$q_p$'. `q_p' is the heat absorbed by the system at constant pressure for increasing the volume from V_1 to V_2. This is so because, -w indicates that work is done by the system. Therefore volume increase against constant pressure is considered.

eqn. becomes qp = û8 3û9

= û+

or

û+ Tp.

`q_p' is the heat absorbed by the system at constant pressure and is considered as the heat content of the system.

Heat effects measured at constant pressure indicate changes in enthalpy of a system and not in changes of internal energy of the system. Using calorimeters operating at constant pressure, the enthalpy change of a process can be measured directly.

Considering a system of gases which are chemically reacting to produce product gases with V_r and V_p as the total volumes of the reactant and product gases respectively, and n_r and n_p as the number of moles of gaseous reactants and products, then using ideal gas law we can write that, at constant temperature and constant pressure,

$$PV_r = nrRT \text{ and } PV_p = npRT.$$

Then considering reactants as initial state and products as final state of the system,

$P(V_p - V_r) = RT\,(n_p - n_r)$

3û9 = ûQ_gRT where,

ûQg refers to the difference in the number of moles of product and reactant

∴ û+ û8 ûQ_gRT

3.7.2 Standard enthalpy changes

The standard enthalpy of a reaction is the enthalpy change for a reaction when all the participating substances (elements and compounds) are present in their standard states.

The standard state of a substance at any specified temperature is its pure form at 1 atm pressure. For example standard state of solid iron at 500 K is pure iron at 500 K and 1 atm. Standard conditions are denoted by adding the superscript 0 to the symbol û+

Similarly, the standard enthalpy changes for combustion, formation, etc. are GHQRWHG E\ û$_c$$H^0$ DQG û$_f$$H^0$etc respectively. Generally the reactants are presented in their standard states during the enthalpy change.

3.8 Thermochemical equations

A balanced chemical equation together with standard conventionsDGRSWHG DQG LQFOXGLQJ WKH YDOXH RI û+ RI WKH UHDFWLRQ LV FDOOHG D thermochemical equation.

The following conventions are necessarily adopted in a thermochemical equation:

(i) The coefficients in a balanced thermochemical equation refer to number of moles of reactants and products involved in the reaction.

(ii) 7KH HQWKDOS\ FKDQJH RI WKH UHDFWLRQ ûrH has unit KJ mol^{-1} and will remain as it is, even if more than one mole of the reactant or product are involved but with only the magnitude changing.

(iii) :KHQ D FKHPLFDO HTXDWLRQ LV UHYHUVHG WKH YDOXH RI û+ LV UHYHUVHG LQ sign with the magnitude remaining the same.

(iv) Physical states of all species are important and must be specified in a WKHUPRFKHPLFDO HTXDWLRQ VLQFH û+ GHSHQGV RQ WKH SKDVHV RI UHDFWDQWV and products.

(v) If the thermochemical equation is multiplied throughout by a number, the enthalpy change is also be multiplied by the same number value.

(vi) 7KH QHJDWLYH VLJQ RI û$_r$H^{0}indicates the reaction to be an exothermic UHDFWLRQ DQG SRVLWLYH VLJQ RI û$_r$H^0 indicates an endothermic type of reaction.

For example, consider the following reaction,

$2H_{2(g)} + O_{2(g)} \rightarrow 2H_2O_{(g)}$û$_r$H^0 =-483.7 KJ.mol^{-1}

$2H_{2(g)} + O_{2(g)} \rightarrow 2H_2O_{(l)}$û$_r$H^0 = -571.1 KJ.mol^{-1}

The above thermochemical equations can be interpreted in several ways.

483.7 KJ given off per mole of the reaction ≡ 483.7

KJ given off per 2 moles of $H_{2(g)}$ consumed≡483.7KJgiven off per mole of $O_{2(g)}$consumed ≡

483.7 KJ given off per 2 moles of water vapour formed

The above equation describes the combustion of H_2 gas to water in a general sense. The first reaction can be considered as the formation reaction of water vapour and the second reaction as the formation of liquid water. Both the reaction refer to constant temperature and pressure.

The negative sign of û+ LQGLFDWHV WKDW LW LV DQ H[RWKHUPLF UHDFWLRQ 7KH reaction which is exothermic in the forward direction is endothermic in the revere direction and vice-versa. This rule applies to both physical and chemical processes.

$2H_2O_{(l)} \rightarrow 2H_{2(g)} + O_{2(g)}$ û$_r$H^0 = +571.1 KJ.mol^{-1}

$2H_2O_{(g)} \rightarrow 2H_{2(g)} + O_{2(g)}$ û+$_r^0$ = +483.7 KJ.mol^{-1}

3.9 Enthalpy of combustion

Generally combustion reactions occur in oxygen atmosphere (excess oxygen) with evolution of heat. These reactions are exothermic in nature. Enthalpy changes of combustion reactions are used in industrial heating and in rocket fuels and in domestic fuels.

(QWKDOS\ FKDQJH RI FRPEXVWLRQ ûcH, of a substance at a given temperature is defined as the enthalpy change of the reaction accompanying the complete combustion of one mole of the substance in presence of excess oxygen at that temperature. The enthalpy changes of combustion of substances in their standard states are known as standard enthalpy change RI FRPEXVWLRQ û$_c$H). These values are useful to experimentally determine the standard enthalpy change of formation of organic compounds.

3.9.1 Bomb calorimeter

Enthalpy changes of combustion of chemical substances are experimentally determined using a bomb calorimeter.

The bomb calorimeter apparatus is shown in Fig.12.3. The inner vessel or the bomb and its cover are made of strong steel. The cover is fitted tightly to the vessel by means of metal lid and screws. A weighed amount of the substance is taken in a platinum cup or boat connected with electrical wires for striking an arc instantly to kindle combustion. The bomb is then tightly closed and pressurized with excess oxygen. The bomb is lowered in water which is placed inside the calorimeter. A stirrer is placed in the space between the wall of the calorimeter and the bomb, so that water can be stirred, uniformly. The reaction is started in the bomb by heating the substance through electrical heating. During burning, the exothermic heat generated inside the bomb raises the temperature of the surrounding water bath. The enthalpy measurement in this case corresponds to the heat of reaction at constant volume. Although the temperature rise is small (only by few degrees), the temperature change can be measured accurately using Beckmann thermometer.

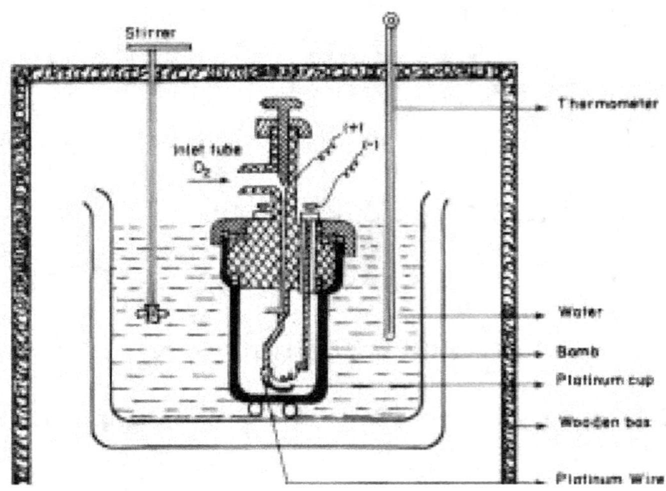

Fig. 3.3 Bomb calorimeter

In a typical bomb calorimeter experiment, a weighed sample of benzoic acid (w) is placed in the bomb which is then filled with excess oxygen and sealed. Ignition is brought about electrically. The rise in temperature û7 LV QRWHG:DWHU HTXLYDOHQW &$_c$) of the calorimeter is known from the standard value of enthalpy of combustion of benzoic acid.

û+$_c{}^oC_6H_5COOH_{(s)}$ = -3227 kJ mol^{-1} w

∴ û+$_c{}^oC_6H_5COOH$ x — = w_c û7 M_2

(where M_2 = mol.weight benzoic acid)

.QRZLQJ &c value, the enthalpy of combustion of any other substance is determined adopting the similar procedure and using the substance in place of benzoic acid. By this experiment, the enthalpy of combustion at constant

YROXPH û+cVol) is known

û+c°(Vol) = wc û7

Enthalpy of combustion at constant pressure of the substance is calculated from the equation,

û+c°(Pr) û+c°(Vol) û$Q_{(g)}$RT

DQG û$Q_{(g)}$ is known from the difference in the number of moles of the products and reactants in the completely balanced equation of combustion of the substance with excess oxygen.

3.10 Enthalpy of neutralization

The **enthalpy change of neutralization** is defined as the enthalpy change accompanied by the complete neutralization of one gram–equivalent amount of a strong acid by a gram-equivalent amount of strong base under fully ionized state in dilute conditions. It is found that the enthalpy of neutralization of a strong acid and a strong base is a constant value equal to -57.32 kJ. This value is independent of the nature of the strong acid and strong base. Strong acids and strong bases exist in the fully ionized form in aqueous solutions as below:

$H_3O^+ + Cl^- + Na^+ + OH^- \rightarrow Na^+ + Cl^- + 2H_2O$ (or)

$H_3O^+_{(aq)} + OH^-_{(aq)} \rightarrow 2H_2O_{(l)}$ûneuH° = -57.32 KJ.

The H^+ ions produced in water by the acid molecules exist as H_3O^+. During the neutralization reaction, water and salt (existing as ions) are produced in solution. Thus, enthalpy change of neutralization is essentially due to enthalpy change per mole of water formed from H_3O^+ and OH^- ions. Therefore, irrespective of the chemical nature, the enthalpy of neutralization of strong acid by strong base is a constant value. At infinite dilutions, complete ionization of acids and bases are ensured and also the inter ionic interactions exist in the lowest extents.

In case of neutralization of a weak acid like acetic acid (CH_3COOH) by a strong base ($NaOH$) or neutralization of weak base (NH_4OH) by a strong acid, two steps are involved. The first step is the ionization of weak acid or weak base since these molecules are only partially ionised. The second step being the neutralization step of H_3O^+ and OH^- ions. Since ionization of weak acids and weak bases in water are endothermic and some energy will be used up in dissociating weak acid and weak base molecules.

Thus, acetic acid with NaOH and ammonium hydroxide with HCl neutralization reactions can be written as,

$$CH_3COOH_{(aq)} + H_2O_2 \rightarrow CH_3COO^-_{(aq)} + H_3O^+_{(aq)}$$

$$Na^+_{(aq)} + H_3O^+_{(aq)} + OH^-_{(aq)} \rightarrow 2H_2O_{(l)} + Na^+_{(aq)} \text{and}$$

$$NH_4OH \rightarrow \quad NH_4^+ + OH^-$$

$$H_3O^+ + Cl^- + OH^- \rightarrow \; 2H_2O + Cl^-.$$

Enthalpy of neutralization of a weak acid or a weak base is equal to -57.32 kJ + enthalpy of ionization of weak acid (or) base. Since enthalpy of ionization of weak acid or base is endothermic it is a positive value, hence enthalpy of neutralization of a weak acid or base will be lower than the neutralization of strong acid and strong base.

The acid with the lowest positive value of heat of ionization will be the strongest acid. Thus formic acid is the strongest and hydrocyanic acid the weakest acid. The trend in decreasing strength of acids is:

û+ IRU LRQL]DWLRQ RI
NH_4OH= +1.50 kCal/g.equiv.

SUMMARY

In this chapter the importance of thermodynamics with a lot of experiments and problems are discussed. The definitions like system, surroundings, intensive and extensive properties, thermodynamic properties are given with brief explanations. The laws of thermodynamics are explained with simple examples. To understand thermodynamics, several problems are given both worked out and practice.

REFERENCES

1. Thermodynamics by Samuel Glasstone.
2. Physical Chemistry by Lewis and Glasstone.
3. Physical Chemistry by Castllan.
4. Physical Chemistry by P.L.Soni.
5. Atkins' Physical Chemistry seventh edition 2002. Oxford University Press
   Page 57

CHAPTER – 4

CHEMICAL EQUILIBRIUM

OBJECTIVES

After Studying this Chapter you will able to:

- *to understand the scopes of chemical equilibrium and to know the extent of completeness of chemical reactions.*
- *to compare and learn about the reversible and irreversible reactions.*
- *to study the dynamic nature of the chemical equilibrium.*
- *to explain the equilibrium existing in physical and chemical changes.*
- *to define law of chemical equilibrium and the equilibrium constant. To express equilibrium constant in terms of concentration and partial pressures and inter relate them.*
- *to deduce expressions for equilibrium constants of homogeneous and heterogeneous chemical equilibria and study suitable examples in each.*

Equilibrium in Chemical Reactions

Consider a chemical reaction between A and B to form products C and D. After allowing sufficient period of time for the reaction, upon analyses, when A and B are absent in the reaction mixture, then the reaction is understood to be complete and only the presence of C and D will be detected. For example, when sodium reacts with water, sodium hydroxide and hydrogen gas are produced, and the reverse reaction to form back the reactants never occurs even when the reaction vessel is a closed one. Reactions when go to completion and never proceed in the reverse direction are called as **irreversible reactions**. The chemical equations of such reactions are represented with a single arrow as $A + B \rightarrow C + D$.

For Example, $2Na + 2H_2O \rightarrow 2NaOH + H_2$

However, even after allowing sufficient period of time for reaction, when the presence of A and B are always detected along with C and D, then such reactions are understood to be never complete.

For example, when H_2 and I_2 are reacted, 2HI is formed. Initially the reaction proceeds to form HI until a certain period of time and with further increase in the reaction time, HI molecules dissociate to produce back H_2 and I_2 in such a way that, the reaction mixture always contain H_2, I_2 and HI for any length of time until external factors like temperature, pressure, catalyst etc. are applied. Reactions which never proceed to completion in both forward and backward direction are called as **Equilibrium reactions**.

The chemical equation of such reactions are represented as,

$$A + B \rightarrow C + D$$

Example, $H_2 + I_2 \rightarrow 2HI$

when both forward and reverse reaction rates are equal, the

concentration of reactants and products do not change with any length of reaction time. Physical transformations of matter like change of solid to liquid states or liquid to vapour states also take place under equilibrium conditions with both the states of matter being present together. For example, at 0˚C, melting ice and freezing water are both present.

4.1 Scope of Chemical Equilibrium

Study of chemical equilibria possesses many scopes. The knowledge on whether the equilibrium lies in favour of reactants or products under certain experimental conditions is useful to increase yields in industrial processes, to establish the exact proton transfer equilibria in aqueous protein solutions. Since small changes in equilibrium concentration of hydrogen ion may result in protein denaturing and cell damage etc. This study is also useful or certain acids, bases and salts in water exist in ionic equilibria which control their use as buffers, colour indicators etc.

4.2 Reversible and Irreversible Reactions

A reaction which can go in the forward and backward direction simultaneously under the same conditions, is called a **reversible reaction**.

If the forward reaction is written as k_f

$A + B \longrightarrow C + D$ then, the reverse reaction is written as

$C + D \xrightarrow{k_r} A + B$. The reversible reaction is represented as

$PCl_{5(s)} \rightleftharpoons PCl_{3(s)} + Cl_{2\,(g)}$

$CaCO_{3(s)} \rightleftharpoons CaO_{(s)} + CO_{2(g)}$.

In a reaction when the product molecules never react to produce back the reactants, then such a reaction is called as **irreversible reaction**. For example,

$NaCl_{(aq)} + AgNO_{3(aq)} \rightarrow NaNO_{3(aq)} + AgCl$

In irreversible reactions only forward reaction takes place and the reaction goes to completion. After the completion, only products exist.

4.3 Nature of Chemical Equilibrium

The occurrence of chemical equilibrium is seen in reversible reactions only. Chemical equilibrium may be defined as the state of a reversible reaction when the two opposing reactions occur at the same rate and the concentration of reactants and products do not change with time. The true equilibrium of a reaction can be attained from both sides.

For X → Y reaction, a = → $\frac{d[x]}{dt}$; b → $\frac{d[Y]}{dt}$

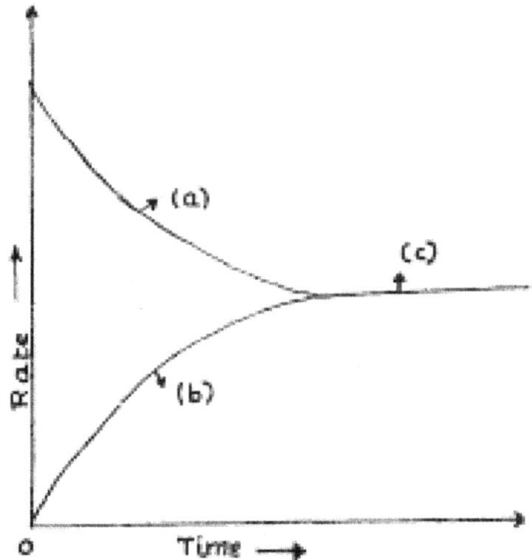

Fig. 4.1.(a) Forward rate (b) reverse rate (c) Equilibriumcondition.

The equilibrium concentrations of reactants and products do not change with time. This is because, since the forward reaction rate equals with backward reaction rate as and when the products are formed, they react back to form the reactants in equal capacity. The equilibrium concentrations of reactants are different from their initial concentrations.

The equilibrium concentrations are represented by square brackets with subscript 'eq' or as [ ]eq. Thus [A]eq denotes the equilibrium concentration of A in moles per litre. In modern practice, the subscript 'eq' is not used.

4.3.1 Dynamic Equilibrium

When a reversible reaction attains equilibrium it appears that the concentrations of individual reactants and that of the products remain constant with time. Apparently, the equilibrium appears as dead (or) as not proceeding. Actually, the reactant molecules are always reacting to form the product molecules. When the product molecules are able to react with themselves under the same experimental condition to form the same amount of reactants simultaneously (at the same time) in an equal rate of the forward reaction, then the process is a ceaseless phenomenon. Thus chemical equilibrium is **dynamic** when the forward and reverse reactions take place **endlessly and simultaneously with equal rates**. Therefore chemical equilibrium is called as dynamic equilibrium.

4.3.2 Characteristics of Chemical Equilibrium

(i) Constancy of concentrations

When a chemical equilibrium is established in a closed vessel at constant temperature, the concentrations of various species like reactants and products remain unchanged.

The reaction mixture consisting of reactants and products at equilibrium is called as equilibrium mixture.

The concentrations of reactants and products at equilibrium are called as equilibrium concentrations.

(ii) **Equilibrium can be initiated from either side.** The state ofequilibrium of a reversible reaction can be arrived at whether we start from reactants or products.

For example, this equilibrium $H_{2(g)} + I_{2(g)} \rightarrow 2HI_{(g)}$ can be achieved whether we start with H_2 and I_2 or with HI.

Equilibrium cannot be attained in an open vessel

Only in a closed vessel, a reaction can be considered to attain equilibrium since no part of reactants or products should escape out. In an open vessel, gaseous reactants or products may escape so that no possibility of attaining equilibrium exists. Equilibrium can be attained when all the reactants and products are in contact with each other.

H_2 HI I_2 → Closed vessel

I_2 H_2 I_2 H_2

$H_2 + I_2 \rightleftharpoons 2HI$

I_2 HI

HI H_2 HI

H_2 I_2

Fig. 4.2 A chemical equilibrium between $H_2 + I_2$ and 2HI

(iii) Catalyst does not alter the equilibrium

When a catalyst is added to the equilibrium system, it speeds up the rates of both forward and reverse reactions to an equal extent. Therefore the equilibrium is not changed but the state of equilibrium is attained earlier.

(iv) The value of equilibrium constant does not depend upon the initial concentration of reactants.

(v) At equilibrium, the free energy change is minimum or zero.

(vi) When temperature is changed, the forward and backward reaction rates are changed and the equilibrium concentrations of reactants and products are changed.

4.3.3 Equilibrium in physical processes

When there is a change in the state of occurrence of matter, then a physical transformation is said to have occurred. The equilibrium concepts are also applicable to physical state transformations of matter.

(i) Solid-liquid equilibria

Here, the solid and the liquid forms of a substance co-exist at characteristic temperature and pressure. At 1 atm and at the melting point of a substance, there is a solid-liquid equilibrium existing. For example, the solid-liquid equilibrium of water at 0°C,

$$\text{water}_{(l)} \rightarrow \text{ice}_{(s)}$$

occurs at 1 atm pressure. Here, both the liquid and ice exist together. Also, at melting point of ice or freezing point of water, the rate of melting of ice equals with rate of freezing of water. With change in pressure the temperature at which this equilibrium onsets changes.

(ii) Liquid-vapour equilibrium

Here the vapour and the liquid forms of a substance exist simultaneously at a characteristic temperature called as boiling point and at 1 atm pressure. For example at 100°C which is the boiling point of water, and 1 atm pressure,

$$\text{Water}_{(l)} \rightarrow \text{Steam}_{(g)}$$

both liquid water and water vapour (steam) exist simultaneously, provided the vapour does not escape.

(iii)Solid-solid equilibrium

When a substance existing in a particular crystalline solid transforms to another crystalline form retaining its solid nature at a characteristic temperature called the transition temperature with both the solid forms coexisting, at 1 atm pressure then it is said to be in solid-solid equilibrium. For example, solid sulphur exhibits equilibrium with rhombic to monoclinic forms at its transition temperature.

$$\text{S(rhombic)} \rightarrow \text{S(monoclinic)}$$

4.3.4 Equilibrium in chemical processes

Chemical equilibrium exists in two types such as homogeneous and heterogeneous equilibria. In a chemical reaction existing in equilibrium, if all the reactants and products are present in the same phase, then a homogeneous equilibria is said to have occurred.

For example,

$N_{2(g)} + 3H_{2(g)} \rightarrow 2NH_{3(g)}$.

Here all the reactants and products exist in gaseous state. This is an example of gas-phase equilibrium.

The chemical equilibrium in which all the reactants and products are in the liquid phase are referred to as liquid equilibria. For example,

$CH_3COOH_{(l)} + C_2H_5OH_{(l)} \rightarrow CH_3COOC_2H_{5(l)} + H_2O_{(l)}$

Both gas phase and liquid phase equilibria are collectively called as homogeneous equilibria.

Heterogeneous equilibrium

In a chemical equilibrium, if the reactants and products are in different phases then heterogeneous equilibrium is said to have occurred.

Examples :

$CaCO_{3(s)} \rightarrow CaO_{(s)} + CO_2$

$3Fe_{(s)} + 4H_2O_{(g)} \rightarrow Fe_3O_4 + 4H_2$

Here, only when the reaction is carried out in closed vessel, the equilibrium state is established.

4.4 Law of chemical equilibrium and equilibrium constant

Law of Mass action

Two Norwegian Chemists, Guldberg and Waage, studied experimentally a large number of equilibrium reactions. In 1864, they postulated a generalization called the **Law of Mass action**. It states that:

"the rate of a chemical reaction is proportional to the active masses of the reactants". By the term `active mass', it is meant the molar concentration i.e., number of moles per litre.

Law of Mass Action based on the Molecular Collision theory

We assume that a chemical reaction occurs as the result of the collisions between the reacting molecules. Although some of these collisions are ineffective, the chemical change produced is proportional to the number of collisions actually taking place. Thus at a fixed temperature the rate of a reaction is determined by the number of collisions between the reactant molecules present in unit volume and hence its concentration, which is generally referred as the active mass.

4.4.1 Equilibrium constant and equilibrium law

Let us consider a general reaction

$$A + B \overset{k_f}{\rightleftharpoons} C + D \quad k_r$$

and let [A], [B], [C] and [D] represent the molar concentrations of A,B,C and D at the equilibrium point. According to the Law of Mass action,

Rate of forward reaction .>$@ >%@ N_f [A] [B]

5DWH RI UHYHUVH UHDFWLRQ .>&@ >'@ N_r [C] [D]

Where k_f and k_r are rate constants for the forward and reverse reactions.

At equilibrium, rate of forward reaction = rate of reverse reaction.

Therefore,

$$K_f[A]\,[B] = K_r[C]\,[D]$$

k_f [C] [D]

or — = ——— ... (1)
kr [A] [B]

At any specific temperature k_f/k_r is a constant since both k_f and k_r are constants. The ratio k_f/k_r is called **equilibrium constant** and is represented by the symbol K_c. The subscript `c' indicates that the value is in terms of concentration of reactants and products. The equation (1) may be written as

$$K_f = \frac{[C]\,[D] \leftarrow \text{Products concentration}}{[A]\,[B] \leftarrow \text{Reactants concentrations}} \quad \ldots (2)$$

Equilibrium constant

This equation is known as the equilibrium constant expression or equilibrium law. Hence [C], [D] [A] and [B] values are the equilibrium concentrations and are equal to equilibrium concentrations.

4.4.2 Equilibrium Constant Expression for a Reaction in General Terms

The general reaction may be written as

$$aA + bB \rightarrow cC + dD.$$

Wherea,b,c and d are numerical quotients of the substance A,B,C and D respectively. The equilibrium constant expression is

$$Kc = \frac{[C]^c\,[D]^d}{[A]^a\,[B]^b} \quad \ldots(3)$$

WhereK_c is the Equilibrium constant. The general definition of the equilibrium constant may thus be stated as:

The product of the equilibrium concentrations of the products divided by the product of the equilibrium concentrations of the reactants, with each concentration term raised to a power equal to the coefficient of the substance in the balanced equation.

For Example

(a) Consider the equilibrium constant expression for the reaction

$N_{2(g)} + 3H_{2(g)} \rightarrow 2NH_{3(g)}$

(1) The equation is already balanced. The numerical quotient of H_2 is 3 and NH_3 is 2.

(2)The concentration of the `product' NH_3 is $[NH_3]^2$.

(3) The product of concentrations of the reactants is $[N_2]\,[H_2]^3$.

Therefore, the equilibrium constant expression is

$$K_c = \frac{[NH_3]^2}{[N_2]\,[H_2]^3}$$

(b) Consider the equilibrium constant expression for the reaction

$N_2O_{5(g)} \rightarrow NO_{2(g)} + O_{2(g)}$

(1) The equation as written is not balanced. Balancing yields

$2N_2O_5 \rightarrow 4NO_2 + O_2$.

(2) The coefficient of the product NO_2 is 4 and of the reactant N_2O_5 is 2.

(3) The product of the concentrations of products is $[NO_2]^4\,[O_2]$.

(4) The concentration of the reactant is $[N_2O_5]^2$

(5) The equilibrium constant expression can be written as

$$K_c = \_[NO_2]^4\_[O_2]\ \_$$

$[N_2O_5]^2$

(c) Consider the equilibrium constant expression of the reaction.

$$CH_{4(g)} + H_2O_{(g)} \rightarrow CO_{(g)} + 3H_{2(g)}$$

(i) Write the product of concentrations of `products' divided by the product (multiplication) of concentrations of `reactants'.

(ii) The concentration of H_2 is to be raised by its coefficient in the balanced equation. Thus, the equilibrium constant expression is:

$$Kc = \frac{[CO]\,[H_2]^3}{[CH_4]\,[H_2O]}$$

4.4.3 Equilibrium Constant Expression for Gaseous Equilibrium

When all the reactants and products are gases, we can formulate the equilibrium constant expression in terms of partial pressures exactly similar to equation (1).

The partial pressure of a gas in the equilibrium mixture is directly proportional to its molar concentration at a given temperature.

Considering equation(1): $K_p = \frac{p_C^c\, p_D^d}{p_A^J\, p_B^K}$

The gaseous equilibrium reaction of SO_2 can be written as follows

$$2SO_{2(g)} + O_{2(g)} \rightarrow 2SO_{3(g)}$$

$$K_p = \frac{(p_{SO3})^2}{(p_{SO2})^2 \cdot p_{O2}} \quad atm^{-1}$$

As an example, K_p value can be calculated from the following data:

The total pressure in the reaction flask is 1 atm and the partial pressures of oxygen, SO_2 and SO_3 at equilibrium are 0.1 atm, 0.57 atm, 0.33 atm respectively.

$$\therefore K_p = \frac{(0.33)^2}{(0.57)^2 \times 0.1} \text{ atm}^{-1}$$

$$= \frac{(0.33)^2}{0.03249} \text{ atm}^{-1}$$

$K_p = 3.3518 \text{ atm}^{-1}$.

In the ammonia formation reaction, the gaseous chemical equilibrium exists as:

$N_{2(g)} + 3H_{2(g)} \rightarrow 2NH_{3(g)}$

where p = partial pressure

Similarly for HI formation in the gaseous state from H_2 and I_2 gases, theequilibrium constant (K_p) can be written as

$H_{2(g)} + I_{2(g)} \rightleftharpoons 2HI_{(g)}$

here K_p has no units.

4.4.4 Degree of dissociation (x)

In the study of dissociation equilibrium, it is easier to derive the equilibrium constant expression in terms of **degree of dissociation (x)**. It is considered as the fraction of total molecules that actually, dissociate into the simpler molecules x has no units. If x is the degree of dissociation then for completely dissociating molecules x = 1.0. For all dissociations involving equilibrium state, x is a fractional value. If x is known, K_c or K_p can be calculated and vice-versa.

Equilibrium constants in terms of degree of dissociation

(i) Formation of HI from H_2 and I_2

The formation of HI from H_2 and I_2 is an example of gaseous homogeneous equilibrium reaction. It can be represented as

$H_{2(g)} + I_{2(g)} \; 2HI_{(g)}$ û+ -10.4 kJ

This equilibrium is an exothermic one.

Let us consider that one mole of H_2 and one mole of I_2 are present initially in a vessel of volume V dm^3. At equilibrium let us assume that x mole of H_2 combines with x mole of I_2 to give 2x moles of HI. The concentrations of H_2, I_2 and HI remaining at equilibrium can be calculated as follows:

According to the law of mass action,

$$K_c = \frac{[HI]^2}{[H_2]\,[I_2]}$$

Substituting the values of equilibrium concentrations in the above equation, we get

If the initial concentration of H_2 and I_2 are equal to a and b moles dm^{-3} respectively, then it can be shown that

$$K_c = \frac{4x^2}{(a-x)(b-x)}$$

Derivation of K_p in terms of x

Let us consider one mole of H_2 and one mole of $I_2$are present initially. At equilibrium, let us assume that x mole of H_2 combines with x mole of I_2 to give 2x moles of HI. Let the total pressure at equilibrium be P atmosphere. The number of moles of H_2, I_2 and HI present at equilibrium can be calculated as follows :

	$H_{2(g)}$	$I_{2(g)}$	$HI_{(g)}$
Initial number of moles	I	I	O
Number of moles reacted	X	x	-
Number of moles remaining at equilibrium	I-x	I-x	2x

∴ The total number of moles at equilibrium	$= 1 - x + 1 - x + 2x = 2$

We know that partial pressure is the product of mole fraction and the total pressure. Mole fraction is the number of moles of that individual component divided by the total number of moles in the mixture. Therefore,

$$p_{H_2} = \frac{1-x}{2}P,\ p_{I_2} = \frac{1-x}{2}P,\ p_{HI} = \frac{2x}{2}P$$

We know that $K_p = \frac{p^2_{HI}}{}$

Substituting the values of partial pressures in the above equation, we get

$$K_p = \frac{\left(\frac{2x}{2}p\right)^2}{\left(\frac{1-x}{2}p\right)\left(\frac{1-x}{2}p\right)} = \frac{4x^2P^2}{4} \times \frac{4}{(1-x)^2 P^2}$$

$$\therefore \quad \boxed{K_p = \frac{4x^2}{(1-x)^2}}$$

we see that K_p and K_e are equal in terms of x values. The influence of various factors on the chemical equilibrium can be explained as below:

(i) Influence of pressure: The expressions for the equilibriumconstants K_c and K_p involve neither the pressure nor volume term. So the equilibrium constants are independent of pressure and volume. Pressure has therefore no effect on the equilibrium.

(ii) Influence of concentration: The addition of either H_2or $I_2$to theequilibrium mixture well increase the value of the denominator in the equation $K_e = [HI]^2/[H_2][I_2]$ and hence tends to decrease the value of K_e. In order to maintain the constancy of K_c, the increase in the denominator value will be compensated by the corresponding increase in the numerator value. In other words, the forward reaction will be favoured and there will be corresponding increase in the concentration of HI.

(iii) Influence of catalyst:A catalyst affects both the forward andreverse reactions to the same extent. So it does not change the relative amounts of reactants and products at equilibrium. The values of K_e and K_p are not affected. However the equilibrium is attained quickly in the presence of a catalyst.

(ii) Dissociation of PCl_5

Phosphorus pentachloride dissociates in gas phase to give PCl_3 and Cl_2. This is an example of gaseous homogeneous equilibrium. It can be represented as

$PCl_{5(g)} \rightleftharpoons PCl_{3(g)} + Cl_{2(g)}$

$K_p = K_c(RT)$

Let us consider that one mole of PCl_5 is present initially in a vessel of volume V dm^3. At equilibrium let x mole dissociates to give x mole of PCl_3 and x mole of Cl_2.

$$K_C = \frac{X_2}{(1-x)V}$$

Derivation of K_p in terms of x

Let us consider that one mole of PCl_5 is present initially. At equilibrium, let us assume that x mole of PCl_5 dissociates to give x mole of PCl_3 and x mole of Cl_2. Let the total pressure at equilibrium be P atmosphere. The number of moles of PCl_5, PCl_3 and Cl_2 present at equilibrium can be given as follows:

	$PCl_{5(g)}$	$PCl_{3(g)}$	$Cl_{2(g)}$
Initial number of moles	1	0	0
Number of moles reacted	X	-	-
Number of moles at equilibrium	1-x	x	x

Total number of moles at equilibrium $= 1 - x + x + x$

$= 1 + x$

We know that partial pressure is the product of mole fraction and the total pressure. Mole fraction is the number of moles of that component divided by the total number of moles in the mixture. Therefore

$$p_{PCl_5} = \frac{1-x}{1+x}P; \quad p_{PCl_3} = \frac{x}{1+x}P; \quad p_{Cl_2} = \frac{x}{1-x}.P$$

we know that $$K_p = \frac{p_{PCl_3} \cdot p_{Cl_2}}{p_{PCl_5}}$$

$$Kp = \frac{x^2P}{1-x^2}$$

When $x << 1$, x^2 value can be neglected when compared to one.

$\therefore K_p \simeq x^2P$

This equation can be used to predict the influence of pressure on this equilibrium.

(i) Influence of pressure: The expression for K_c contains the volumeterm and the expression for K_p contains the pressure term. Therefore this equilibrium is affected by the total pressure. According to the above equation, increase in the value of P will tend to increase the value of K_p. But K_p is a constant at constant temperature. Therefore, in order to maintain the constancy of K_p the value of x should decrease. Thus, increase in total pressure favours the reverse reaction and decreases the value of x.

(ii) Influence of concentration: Increase in the concentration of PCl_5favours the forward reaction, while increase in the concentration of either PCl_3 or Cl_2favours the reverse reaction. Increase in the concentration of a substance in a reversible reaction will favour the reaction in that direction in which the substance is used up.

(iii) Influence of catalyst:A catalyst will affect the rates of theforward and reverse reactions to the same extent. It does not change either the amount of reactants and products present at equilibrium or the numerical value of K_p or K_p. However, the equilibrium is obtained quickly in the presence of a catalyst.

4.4.5 Characteristics of Equilibrium constant

(i) The K_{eq} (K_c or K_p) values do not depend on the initial concentrations of reactants but depend only on the equilibrium concentration values.

(ii) When K_{eq} values are greater than unity, the equilibrium is favourable towards product formations and vice-versa.

(iii) K_{eq} values do not change in presence of catalyst catalyst only speeds up the forward and backward reactions.

(iv) Temperature changes on the chemical equilibrium changes the K_{eq} values. For exothermic equilibrium reactions increase in temperature, lowers the K_{eq} values and for endothermic equilibrium, increase in temperature increases the K_{eq} values. Generally, when T is raised, the equilibrium shifts in a direction in which heat is absorbed.

(v) Pressure changes on gaseous equilibrium alter the K_p values. For dissociation equilibrium, increase in pressure lowers the K_p values, while for association equilibrium (number of product molecules < number of reactant molecules) increase in pressure increases the K_p values. When the number of product and reactant molecules are equal, there is no pressure effect.

4.5 Heterogeneous equilibria

The chemical equilibrium in which the reactants and products are not in the same phases are called **heterogeneous equilibrium**. An example of heterogeneous equilibrium can be the decomposition of calcium carbonate which upon heating forms calcium oxide and carbondioxide under equilibrium conditions. When the reaction is carried out in a closed vessel, the following heterogeneous equilibrium is established:

$CaCO_3 \rightleftharpoons CaO_{(s)} + CO_{2(g)}$

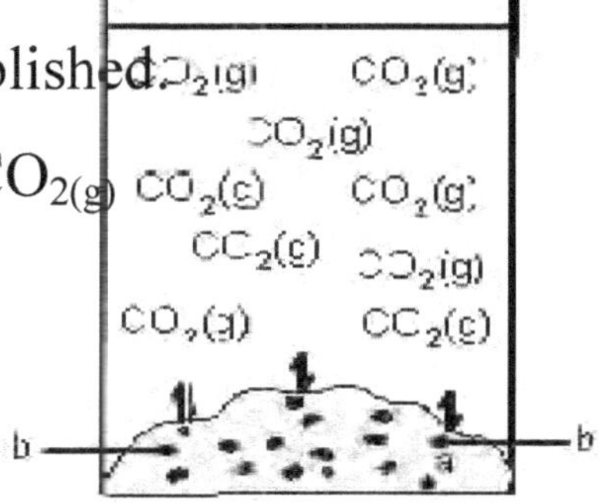

Fig. 13.3 Heterogeneous equilibrium

a = $CaCO_3$Solid; b = CaO Solid

The equilibrium constant expression for $CaCO_3$ dissociation can be written as

$$K = \frac{[CO_2][CaO]}{[CaCO_3]}$$

But CaO and $CaCO_3$ are pure solids. The activity or concentration of pure solids is unity.

Thus $K_c = [CO_2]$

in terms of partial pressures,

$K_p = pco_2$, where Pco_2 is the pressure of CO_2 alone in equilibrium. There are many examples ofheterogeneous chemical equilibria with K_p and K_c values different depending on the number of product and reactant molecules.

(i) The equilibrium constant expression for decomposition of liquid water would be

$2H_2O_{(l)} \rightleftharpoons 2H_{2(g)} + O_{2(g)}$

$$K = \frac{[H_2]^2\,[O_2]}{[H_2O]^2}$$ but activity of liquid $H_2O = 1.0$

$\therefore K_c = [H_2]^2\,[O_2]$

and$K_p = (p_{H2})^2\,(p_{O2})$ p = partial pressure.

(ii) Consider the equilibrium reaction of decomposition of NH_4Cl.

$NH_4Cl_{(s)} \rightleftharpoons NH_{3(g)} + HCl_{(g)}$

$K_c = [NH_3] [HCl]$ because $[NH_4Cl_{(s)}] = 1.0$

$K_p = P_{NH3}P_{HCl}$ P = partial pressure

(iii) Consider the hydrogen gas evolution equilibrium such as

$3Fe_{(s)} + 4H_2O_{(g)} \rightarrow Fe_3O_4 + 4H_{2(g)}$

$$K_c = \frac{[H_2]^4}{[H_2O]^4} \text{ and } k_p = \frac{(p_{H2})^2}{(p_{H2O})^4}$$

since activities of $Fe_{(g)}$ and $Fe_3O_{4(s)}$ are 1.0

SUMMARY

- When the products of a chemical reaction do not react back to give the reactants, then the reaction is called as irreversible reaction.
- In reversible reactions in a closed system, when the rate of forward reaction equals the backward reaction, equilibrium state is reached. The equilibrium concentrations do not change with time.
- For a general equilibrium reaction,

  $aA + bB \rightleftharpoons cC + dD,$

  the equilibrium content, K_{eq} is given by $[C]^c [D]^d / [A]^a [B]^b$.
- For gaseous reaction K_{eq} can be expressed in partial pressures also which is `K_p' value.

  $K_p = K_c (RT)^{ûQ}$ ZKHUH ûQ LV WKH difference in the number of molecules of products and reactants in the equilibrium.
- K_{eq}value depend on Temperature, pressure, and equilibria concentrations and does not depend on catalyst and initial concentrations.
- Concepts of chemical equilibrium also applies to physical

equilibria like solid to liquid, liquid to vapour and solid to solid physical state transformations which take place at constant temperature.

- When the places of the reactants and products in the equilibrium reaction are different like in solid (or) in liquid state then heterogeneous equilibrium is formed. In the equilibrium constant expression, the activities of pure solid and pure liquid form of reactants (or) products are taken as unity.

REFERENCES

1. Physical Chemistry by Maran and Prutton.
2. Physical Chemistry by Lewis and Glasstone.
3. Physical Chemistry by P.W. Atkins.

CHAPTER – 5

CHEMICAL KINETICS

OBJECTIVES

- *To study the scope of chemical kinetics as to know the mechanism of reactions for maximum yield and rate in industrial processes.*
- *To define rate of chemical reactions and to deduce the rate constant of the overall process. Also, the factors affecting the rate and the rate constant are studied. To write the differential rate equations for simple reactions.*
- *To study the effects of nature of the reactant, temperature, concentration of the reactants, presence of catalyst surface area of reactants and irradiation on the overall rate of the processes are to be learnt.*
- *To write the rate law and to understand rate constant, order of the reaction. Different units of the rate constant for various orders of the reaction are to be deduced.*
- *Classification of rates of reactions based on order of the reaction is studied with suitable examples. Hence, zero, first, second, pseudo first order, fractional order, third order reactions are studied.*

5.1 Scope of chemical kinetics

Chemical kinetics is the study of the rates and the mechanism of chemical reactions. Commonly the measure of how fast the products are formed and the reactants consumed is given by the rate values.

The study of chemical kinetics has been highly useful in determining the factors that influence the rate, maximum yield and conversion in industrial processes. The mechanism or the sequence of steps by which the reaction occurs can be known. It is also useful in selecting the optimum conditions for maximum rate and yield of the chemical process.

5.1.1 Rate of chemical reactions

The rate of a reaction tells us how fast the reaction occurs. Let us consider a simple reaction.

$A + B \rightarrow C + D$

As the reaction proceeds, the concentration of the reactant A and B decreases with time and the concentration of the products C + D increase with time simultaneously. The rate of the reaction is defined as the change in the concentration of any reactant or product in the reaction per unit time.

For the above reaction,

Rate of the reaction

= Rate of disappearance of A

= Rate of disappearance of B

= Rate of appearance of C

= Rate of appearance of D

During the reaction, changes in the concentration is

infinitesimally small even for small changes in time when considered in seconds. Therefore differential form of rate expression is adopted. The negative sign shows the concentration decrease trend and the positive sign shows the concentration increase trend.

$$\text{Rate} = \frac{\text{concentration change}}{\text{time taken}} = \frac{-\Delta[A]}{\Delta t}$$

$$\therefore = \frac{-d[A]}{dt} = \frac{-d[B]}{dt} = \frac{+d[C]}{Dt} = \frac{+d[D]}{dt}$$

For a general balanced reaction, written with stoichiometric like x,y, for the reactant and l,m for the product, such as

$xA + yB \rightarrow lC + mD$. The reaction rate is

$$\text{Rate} = -\frac{1}{x}\frac{d[A]}{dt} = -\frac{1}{y}\frac{d[B]}{dt} = +\frac{1}{l}\frac{d[C]}{dt}$$

$$= +\frac{1}{m}\frac{d[D]}{dt}$$

For example: In the reaction,

$H_2 + Br_2 \rightarrow 2HBr$

The overall rate of the reaction is given by

$$\text{Rate} = \frac{-d[H_2]}{Dt} = \frac{-d[Br_2]}{dt} = \frac{1}{2}\frac{d[HBr]}{dt}$$

Consider the reaction, $2NO + 2H_2 \rightarrow N_2 + 2H_2O$

$$\text{Rate} = -\frac{1}{2}\frac{d[NO]}{dt} = \frac{-1}{2}\frac{d[H_2]}{dt} = \frac{c[N_2]}{dt} = \frac{1}{2}\frac{d[H_2O]}{dt}$$

Units of Rate

Reaction rate has units of concentration divided by time. Since concentration is expressed in mol lit^{-1} or mol dm^{-3} the unit of the reaction rate is mol lit^{-1} s^{-1} or mol dm^{-3} s^{-1}.

5.1.2 Factors influencing reaction rates

There are number of factors which influence the rate of the reaction. These are :

(i) Nature of the reactants and products
(ii) Concentration of the reacting species
(iii) Temperature of the system
(iv) Presence of catalyst
(v) Surface area of reactants
(vi) Exposure to radiation

(i) Effect of nature of the reactant and product

Each reactant reacts with its own rate. Changing the chemical nature of any reacting species will change the rate of the reaction. For example, in halogenation reactions, the reactions involving iodine is found to be slower than those involving chlorine.

In case of products, some of them are capable of reacting back to form reactants or some other kind of products. In such cases, the overall rate will be altered depending on the reactivity of the products.

(ii) Effect of reacting species

As the initial concentration of the reactants increase in the reaction mixture, the number of reacting molecules will increase. Since the chemical reaction occurs when the reacting species come close together and collide, the collisions are more frequent when the concentrations are higher. This effect increases the reaction rate.

(iii) Effect of temperature

Increase in temperature of the system increases the rate of the reaction. This is because, as the temperature increases the kinetic energy of the molecules increases, which increases the number of collisions between the molecules. Therefore the overall rate of the reaction increases. This condition is valid only for endothermic reaction. For exothermic reaction the overall rate decreases with increasing temperature.

(iv) Effect of presence of catalyst

A catalyst is a substance that alters the rate of a chemical reaction, while concentration of catalyst remaining the same before and after the reaction. The addition of catalyst generally increases the rate of the reaction at a given temperature. Also, catalyst is specific for a given reaction.

(v) Effect of surface area of reactants

In case of reactions involving solid reactants and in case of heterogeneous reactions, surface area of the reactant has an important role. As the particle size decreases surface area increases for the same mass. More number of molecules at the surface will be exposed to the reaction conditions such that the rate of the reaction increases. Thus the reactants in the powdered form (or) in smaller particles react rapidly than when present in larger particles.

(vi) Effect of radiation

Rates of certain reactions are increased by absorption of photons of energy. Such reactions are known as photochemical reactions. For example, H_2 and Cl_2 react only in the presence of light. With increase in the intensity of the light (or) radiation, the product yield increases. For photosynthesis light radiation is essential and the process does not proceed in the absence of light.

5.1.3 Rate law

According to concepts of chemical kinetics, the rate of the reaction is proportional to the product of the initial concentration of all the reactants with each reactant concentration raised to certain exponential powers.

Consider a general reaction $pA + qB \rightarrow cC + dD$.

The rate law is given by the expression,

$$\text{Rate} = [A]^p\,[B]^q$$

$$\therefore\ \text{Rate} = k[A]^p\,[B]^q$$

where k is proportionality constant also known as the rate constant or velocity constant of the reactions. p and q represent the order of the reaction with respect to A and B. The values of k, p and q are experimentally determined for a given reaction. Values of p and q need not be same as the stoichiometric coefficients of the reaction.

Rate constant

In the above general equation k represents the rate constant. Rate constant or velocity constant (or) specific reaction rate is defined as the rate of the reaction when the concentration of each of the reactants is unity in the reaction.

When concentration of A and B is unity then, the rate constant is equal to the rate of the reaction. When the temperature of the reaction mixture changes, the value of rate constant changes.

5.1.4 Order of the reaction

Order of a reaction is defined as the sum of the exponential powers to which each concentration is raised in the rate expression. For example, if the overall rate is given by the expression

Rate = $k[A]^p [B]^q$

Then, the overall order of the reaction is (p+q). The order with respect to A is p. The order with respect to B is q. If p=1; q=0 and vice versa, the order of the reaction is 1, and the reaction is called first order. If p=1, q=1, the order of the reaction is 2 and the reaction is called second order and so on.

A zero order reaction is one where the reaction rate does not depend upon the concentration of the reactant. In this type of reaction, the rate constant is equal to the rate of the reaction.

5.1.5 Unit of rate constant

In general, rate expression for the reaction,

$pA + qB \rightarrow cC + dD$

Rate =k $[A]^p [B]^q$

$$k = \frac{Rate}{[A]^p[B]^q}$$

The unit for the rate constant `k' depends upon the rate of the reaction, the concentration of the reactants and the order of the reaction.

In the case of the first order reaction.

$$k = \frac{Rate}{[A]^1 [B]^0}$$

$$k = \frac{moldm^{-3}\ sec^{-1}}{moldm^{-3}}$$

Unit of

$k = sec^{-1}$ for first order reaction.

Similarly unit of $k = mol^{-1}\ dm^{3}\ sec^{-1}$ for second order reaction

unit of $k = mol^{-(n-1)}dm^{3(n-1)}\ sec^{-1}$ for n^{th} order reaction

Following are the important differences between rate and rate constant of a reaction

Rate of reaction	Rate constant of reaction
1. It represents the speed at which the reactants are converted into products at any instant.	1. It is the constant of proportionality in the rate law expression.
2. At any instant of time, the rate depends upon the concentration of reactants at that instant.	2. It refers to the rate of a reaction at the specific point when concentration of every reacting species is unity.
3. It decreases as the reaction proceeds.	3. It is constant and does not depend on the progress of the reaction.
4. Rate of rate determining step determines overall rate value.	4. It is an experimental value. It does not depend on the rate determining step.

5.2 Molecularity of the reaction

Molecularity is defined as the number of atoms or molecules taking part in an elementary step leading to a chemical reaction. The overall chemical reaction may consist of many elementary steps. Each elementary reaction has its own molecularity which is equal to number of atoms or molecules participating in it. If the reaction takes place in more than one step there is no molecularity for the overall reaction. However molecularity and order are identical for elementary reaction (one step).

There are many differences between the concepts of order and molecularity.

5.2.1 Rate determining step

Most of the chemical reactions occur by multistep reactions. In the sequence of steps it is found that one of the steps is considerably slower than the others. The overall rate of the reaction cannot be lower in value than the rate of the slowest step. Thus in a multistep reaction the experimentallydetermined rate corresponds to the rate of

Order of a reaction	Molecularity of a reaction
1. It is the sum of powers raised on concentration terms in the rate expression.	1. It is the number of molecules of reactants taking part in elementary step of a reaction.
2. Order of a reaction is an experimental value, derived from rate expression.	2. It is a theoretical concept.
3. Order of a reaction can be zero, fractional or integer.	3. Molecularity can neither be zero nor fractional.
4. Order of a reaction may have negative value.	4. Molecularity can never be negative.
5. It is assigned for overall reaction.	5. It is assigned for each elementary step of mechanism.
6. It depends upon pressure, temperature and concentration (for pseudo order)	6. It is independent of pressure and temperature.

the slowest step. Thus the step which has the lowest rate value among the other steps of the reaction is called as the rate determining step (or) rate limiting step.

Consider the reaction,

$2A + B \rightarrow C + D$ going by two steps like,

$A + B \xrightarrow{k_1} C + Z$ - (1) step (slow)

$Z + A \xrightarrow{k_2} D$ - (2) step (fast)

$2A + B \rightarrow C + D$

Here, the overall rate of the reaction corresponds to the rate of the first step which is the slow step and thus, the first step is called as the rate determining step of the reaction. In the above reaction, the rate of the reaction depends upon the rate constant k1 only. The rate of 2^{nd}step doesn't contribute experimentally determined overall rate of the reaction.

5.2 Classification of rates based on the order of the reaction

The rate law for a reaction must be determined by experiment. Usually the order of the reaction determined experimentally does not coincide with the stoichiometric coefficients of the reactants or products in the balanced chemical equation. Each reaction proceeds by a rate value determined by the rate constant and initial concentrations of the reacting species. Rate constant values differ for different `order' reactions even if concentrations are maintained the same. Therefore chemical reactions are classified according to its rate of chemical transformation which inturn depend on the order of the reaction. Let us consider a general rate equation such as

$$\text{rate} = k[A]^p [B]^q$$

Total order is p + q and order with respect to A is p and with respect to B in q respectively.

Zero order reaction

A reactant whose concentration does not affect the reaction rate is called as zero order reaction,

rate law is,

$rate = k[A]^0$ (or)

$$\frac{-d[A]}{dt} = k \quad \text{or} \quad k = \frac{[A]_0-[A]t}{t}$$

Examples of zero order reaction is h

$$H_{2(g)} + Cl_{2(g)} \rightleftharpoons 2HCl_{(g)}$$

The first order reaction

when aqueous solution of NH_4NO_2 is warmed it decomposes rapidly to H_2O and N_2.

$$NH_4NO_2 \rightarrow 2H_2O + N_2$$

This reaction goes by first order manner rate constant k is given by

$$K = \frac{2.303}{t} \log \frac{V_\infty}{V_\infty - V_t} \sec^{-1}$$

V_∞ and V_t are volume of N_2 collected at room temperature and 1 atm when a fixed amount of NH_4NO_2 decomposes at $t = \infty$ (after completion of reaction) and at any time `t'.

Second order reaction

A reaction is said to be second order if its reaction rate is determined by the variation of two concentration terms or square of a single concentration term.

Third order reactions

A reaction is said to be third order if its rate is determined by the variation of three concentration terms.

Example

$2NO + Cl_2 \rightarrow 2NOCl$

$2FeCl_3 + SnCl_2 \rightarrow 2FeCl_2 - SnCl_4$

$3A \rightarrow Products$

Rate = $k[A]^3$ (or)

$$k = \frac{1}{2t}\left[\frac{1}{[A]_t^2} - \frac{1}{[A]_o^2}\right] \text{ (or) } k = \frac{1}{2t}\left[\frac{x(2a-x)}{a^2(a-x)^2}\right] \text{lit}^2 \text{ mol}^{-2} \text{ sec}^{-1}$$

A second (or) third (or) any other high order reaction can be experimentally followed in an easy way by reducing the overall order to first order type by adopting pseudo order conditions. In this method, excluding the concentration of one of reactants, concentrations of all other reactant are kept in excess (at least 10 times) of the concentration of one of the reactant whose concentrations are to be varied to study the changes in the rate.

Example: Thermal decomposition of acetaldehyde.

$$CH_3CHO \longrightarrow CH_4 + CO$$

Order = 1.5. Here a chain mechanism has been proposed. Polymerization reactions also show fractional orders.

SUMMARY

- Basic concepts of chemical kinetics used in writing the rate law of a general reaction along with order and rate constant units and various expressions are understood.
- Identification of rate determining step examples of reactions with measurable rates are studied. Decomposition of N_2O_5 reaction with various forms of rate expressions are studied.
- Order and molecularity are differentiated and various experimental methods of determination of order of the reaction was understood.
- Classifications of rates based on the order of the reaction are understood with suitable examples each of the zero, first, second,

pseudo first, third and fractional order reactions.

Order	**Unit of k**
Zero	mol litre^{-1} time^{-1}
I	time^{-1}
II	litre mol^{-1} time^{-1}
III	litre^{2} mol^{-2} time^{-1}
nth	litre$^{(n-1)}$ mol$^{(1-n)}$ s^{-1}

REFERENCES

1. Physical Chemistry by Lewis and Glasstone.

CHAPTER – 6

THE SOLID STATE

OBJECTIVES

After studying this chapter, you will be able to

- *Definition of a crystalline solid.*
- *Difference between crystalline and an amorphous materials.*
- *Definition of a unit cell.*
- *Study of Sodium Chloride and Cesium Chloride unit cells.*
- *Definition of Miller Indices.*
- *Learn to identify the important planes in a cubic system in terms of Miller indices.*
- *To recognizes different types of cubic crystal system.*

6.1 Crystalline solids

Some solids, like sodium chloride, sulphur and sugar, besides being incompressible and rigid, have also characteristic geometrical forms. In these solids the atoms or molecules are arranged in a very regular and orderly fashion in a three dimensional pattern. Such substances are called **crystalline solid.**

The X-ray diffraction studies reveal that their ultimate particles (viz., molecules, atoms or ions) are arranged in a definite pattern throughout the entire three-dimensional net-work of a crystal. This definite and ordered arrangement of molecules, atoms or ions (as the case may be) extends over a large distance. This is termed as **long-range order.**

The outstanding characteristics of a crystalline solid are its sharp melting point. Crystalline solids are **anisotropic** since they exhibit different physical properties in all directions e.g., the electrical and thermal conductivities are different in different directions.

Amorphous solids

There is another category of solids such as glass, rubber and plastics, which possess properties of incompressibility and rigidity to a certain extent but do not have definite geometrical forms. Such substances are called **amorphous solids**

Amorphous solids (from the Greek words for 'without form') neither have ordered arrangement nor sharp melting point like crystals but when heated, they become pliable until they assume the properties usually related to liquids. These solids lack well-defined faces and shapes. Many amorphous solids are mixture of molecules that do not stick together well. Most others are composed of large complicated molecules. Amorphous solids are therefore regarded as super cooled liquids with high material becomes rigid but there the forces of attraction holding the molecules together are so great that the material becomes rigid but there is no regularity of structure. Thus, amorphous solids do not melt at specific temperatures. Instead they soften over a temperature range as intermolecular forces of various strengths are overcome.

Amorphous solids are **isotropic** as they exhibit same physical properties in all the directions.

Difference between Crystalline and Amorphous Solids. Crystallineand amorphous solids differ from one another in the following respects

1. Characteristic geometry

A crystalline solid has a definite and regular geometry due to definite and orderly arrangement of molecules or atoms in three-dimensional space. An amorphous solid, on the other hand, does not have any pattern of arrangement of molecules or atoms and, therefore, does not have any define geometrical shape. It has been found that even if some orderly arrangement of molecules or atoms exists in a few amorphous solids, it does not extend more than a few Angstrom units. Thus unlike crystalline solids, amorphous solids do not have a long range order.

2. Melting points

As a solid is heated, it's molecular vibrations increase and ultimately becomes so great that molecules break away from their fixed positions. They now begin to move more freely and have rotational motion as well. The solid now changes into liquid state. The temperature at which this occurs is known as the melting point.

A crystalline substance has a sharp melting point, i.e., it changes abruptly into liquid state. An amorphous substance, on the contrary, does not have a sharp melting point. For example, if glass is heated gradually, it softens and starts to flow without undergoing a definite and abrupt change into liquid state. The amorphous solids are, therefore, regarded as liquids at all temperatures. There is some justification for this view because it is known form X-ray examination that amorphous substance do not have well-ordered molecular or atomic arrangements. Strictly speaking, solid state refers to crystalline state, i.e., only a crystalline material can be considered to be a true solid.

3. Isotropy and Anisotropy

Amorphous substances differ from crystalline solids and resemble liquids in another important respect. The properties such as electrical conductivity thermal conductivity, mechanical strength and refractive index are the same in all directions. Amorphous substances are, therefore, said to be isotropic. Liquids and gases are also isotropic. Crystalline solids, on the other hand, are anisotropic, *i.e., their physical properties are different in different directions.* For example, the velocity of lightpassing through a crystal varies with the direction in which it is measured. Thus, a ray of light entering such a crystal may split up into two components each following a different path and travelling with a different velocity. This phenomenon is known as **double refraction**. Thus, anisotropy in itself is a strong evidence for the existence of ordered molecular arrangements in such materials. This can be shown on reference to Fig. 8.1 in which a simple two-dimensional arrangement of only two different kinds of atoms is depicted.

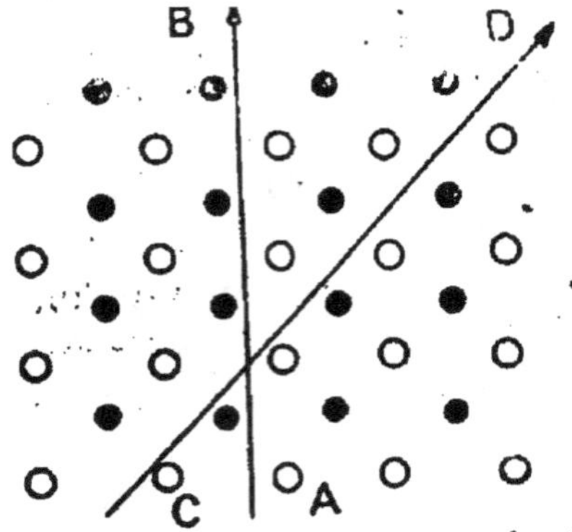

Fig. 6.1 Anisotropic behaviour of crystals

If the properties are measured along the direction indicated by the slanting line CD, they will be different from those measured in the direction indicated by the vertical line AB. The reason is that while in the first case, each row is made up of alternate type of atoms, in the second case, each row is made up of one type of atoms only. In amorphous solids as well as in liquids and gases, atoms or molecules are arranged at random and in a disorderly manner and, therefore, all directions are identical and all properties are alike in all directions.

Size and shape of crystals

Several naturally occurring solids have definite crystalline shapes, which can be recognized easily. There are many other solid materials, which occur as powders or agglomerates of fine particles and appear to be amorphous. But when an individual particle is examined under a microscope, it is also seen to have a definite crystalline shape. Such solids, in which the crystals are so small that can be recognized only under a powerful microscope, are said to be

microcrystalline. The size of a crystal depends on the rate at which it is formed: the slower the rate the bigger the crystal. This is because time is needed by the atoms or molecules to find their proper positions in the crystal structure. Thus, large transparent crystals of sodium chloride, silver chloride, lithium chloride, etc. can be prepared by melting these salts and allowing them to cool very slowly at a uniform rate. It is for this reason that crystals of most of the minerals formed by geological processes are often very large.

Crystal possess the following characteristic feature:

i) Faces: Crystals are bound by plane faces. The surfaces usually planner and arranged on a definite plane (as a result of internal geometry), which bind crystals are called faces.

Faces are of two types:

Like: A crystal having all faces alike e.g. Fluorspar.

Unlike: A crystal having all faces not alike e.g. Galena.

ii) Form: All the faces corresponding to a crystal are said to constitute a form.

iii) Edges: The intersection of two adjacent faces gives rise to the formation of edge.

iv) Interfacial Angle: The angle between the normals to the two intersecting faces is called interfacial angle.

Although the size of the faces or even faces of the crystals of the same substance may vary widely with conditions of formation, etc., yet the interfacial angles for any two corresponding faces of the crystals remain invariably the same throughout.

Although the external shape is different yet the interfacial angles are same. The measurement of interfacial angles in crystals is, therefore, important in the study of crystals. The subject is known as **crystallography.**

6.2 Unit Cell

Crystals are built up of a regular arrangement of atoms or ions in three dimensions. *The smallest structure of which the crystalline solid (orcrystal) is built by its repetition in three dimensions is called as unit cell.*A unit cell may be considered as the brick of a wall depends upon the shape of brick, the shape of crystal also depends upon the shape of unit cell. Therefore, a **unit** cell is the fundamental elementary pattern of a crystalline solid. The characterization of the crystal involves the identification of its unit cell.

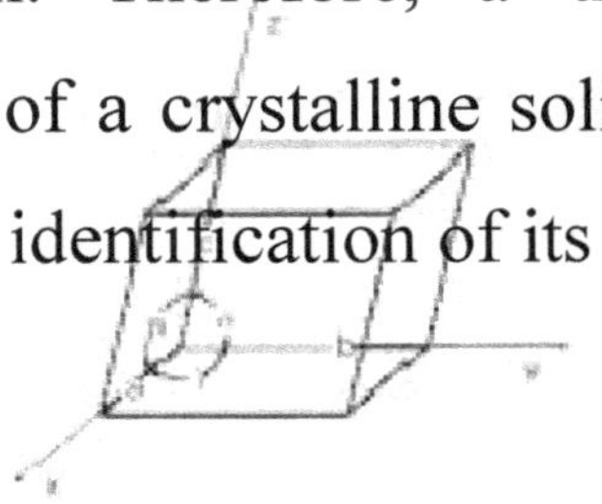

Characteristic parameters of unit cell

1. Crystallographic axes: The lines drawn parallel to the lines of intersection of any three faces of the unit cell which do not lie in the same plane are called crystallographic axes.
2. Interfacial angles: The angles between the three crystallographic axes are known as interfacial angles.
3. Primitives: The three sides a, b and c (as shown in figure) of a

unit cell are known as primitives or characteristic intercepts.

Crystallographic axes: OX, OY, OZ Interfacial angles: α, β, γ Primitives (distances): a, b, c

The unit cell is characterized by the distances a, b and c and angles α, β,&γ.

The size (edge length) of a unit cell depends on the size of the atoms or ions and their arrangement. Because a unit cell is representative of the entire structure, the ratio of ions in the unit cell is the same as the ratio in the overall structure.

There are seven classes of unit cells. *1. Cubic , 2. Triclinic, 3. Monoclinic, 4. Orthorhombic, 5. Tetragonal, 6. Hexagonal and 7. Rhombohedral.*

Packing arrangement variations exist in each of the classes, yet here we will only explore the cubic system because it is the simplest.

Cubic unit cells

Three types of cubic unit cells are discussed in this chapter. Each unit cell is defined by one type of atom. Consequently, whichever atom you choose when defining that unit cell is the ***only*** atom used in defining the unit cell. (Ignore all other types)

Simple cubic

The most simple unit cell is known as a simple cubic unit cell. This is where one atom occupies each of the eight corners of a cube. The distance from atom to atom along the lattice is the same in every direction, and the angle between each of axes is 90°.

Body-centered cubic

The next unit cell is known as the body-centered cubic. In this form of crystal, there is an atom at each corner of the unit cell and also there is an additional atom in the center of the cube. This packing can fit more atoms into less space than the simple cubic unit cell.

Face-centered cubic

A slightly more tightly packed unit cell is the face-centered cubic unit cell. In this form of crystal, there is an atom at each corner of the unit cell. And there is an atom at the center of each of the six faces of the cubic unit cell. This crystal packing form has an even higher density than the body-centered cubic unit cell.

Atoms or ions are shared between adjacent unit cells. The lattice position of the atom or ion determines the number of unit cells involved in the share. There are four different lattice positions an atom or ion can occupy.

Body: Not shared

Face: Shared by two unit cells

Edge: Shared by four unit cells

Corner: Shared by eight unit cells.

Sodium chloride crystal

Space lattice of sodium chloride is known to consist of a face-centered cubic lattice of Na^+ ions interlocked with a similar lattice of Cl^-ions. A unit cell of this combined lattice is shown in figure. This unit cell repeats itself in three dimensions throughout the entire crystal. The yellow spheres indicate chloride ions and red spheres represent sodium ions. The lattices are constituted entirely by ions are known as **ioniclattices**. All electrovalent compounds show such

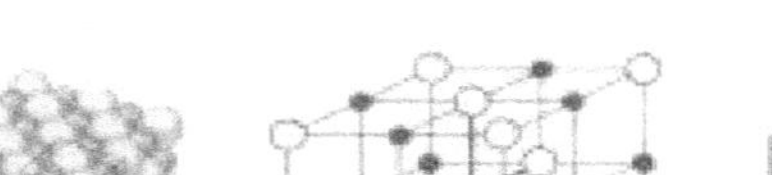

lattices.

There are four units of NaCl in each unit cube with atoms in the positions

Cl: 0 0 0; ½ ½ 0; ½ 0 ½; 0½ ½ ;

Na: ½ ½ ½; 0 0 ½; 0 ½ 0; ½ 0 0;

As will be seen figure, the unit cell of sodium chloride consists of 14 chloride ions and 13 sodium ions. Each chloride ion is surrounded by 6 sodium ions and similarly, each sodium ion is surrounded by 6 chloride ions.

Notice that the particles at corners, edges and faces do not lie wholly within the unit cell. Instead these particles are shared by other unit cells. A particle at a corner is shared by eight unit cells, one at the centre of face is shared by two and one at the edge is shared by four. The unit cell of sodium chloride has 4 sodium ions and 4 chloride ions as shown below;

No. of Sodium ions

= 12 (At edge centers) x (1/4) + 1 (At body center) x_1

1 1 1 1

= HF + E+) + &

= 12 x ¼ + 1x 1 **=3 + 1 =4**

No. of Chloride ions

= 8 (At corners) x (1/8) + 6 (At face centre) x (1/2)

= 8 x (1/8) + 6 x (1/2) **=1+ 3** = **4**

Thus, number of NaCl units per unit cell is 4.

The sodium chloride structure is also called rock-salt structure.

Representative crystals having the NaCl arrangements includes: LiH, NaI, KCl, RbF, RbI, PbS etc.

Cesium Chloride structure

The cesium chloride, CsCl, structure has body-centered cubic system and is shown in figure. The body-centered cubic arrangement of atoms is not a close packed structure. There is one molecule per primitive cell, with atoms at the corners (000) and body-centered positions 1/2 1/2 1/2 of the simple cubic space lattice.

For example consider plane LMN in the crystal shown in Fig. 6.2.

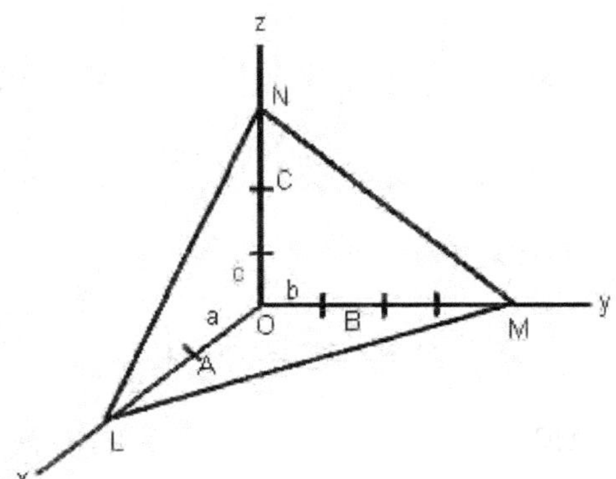

Fig. 6.2 The intercepts of a crystallographic plane

This plane has intercepts OL, OM and ON along the x-, y- and z-axes at distances 2a, 4b and 3c respectively, when OA = a, OB = b and OC = c are the chosen unit distances along the three coordinates. These intercepts are in the ratio of 2a : 4b : 3c wherein 2, 4, 3 are simple integral whole numbers.

The coefficients of a, b and c (2, 4 and 3 in this case) are known as the Weiss indices of a plane. It may be borne in mind that the Weiss indices are not always simple integral whole numbers as in this case. They may have fractional values as well as infinity (an indefinite quantity). Weiss indices are, therefore, rather awkward in use and have consequently been replaced by miller indices. Taking the reciprocals of Weiss indices and multiplying throughout by the smallest number in order to make all reciprocals as integers obtain the **Miller indices** of a plane. The Miller indices for a particular family of planes are usually written **(h, k, l)** where h, k and l are positive or negative integers or zero.

Consider a plane which in Weiss notation is given by 2 a: 4b: 3c. Taking reciprocals of coefficients of a, b and c, we get the ratio 1/2, 1/4, 1/3. Multiplying by 12 in order to convert them into whole numbers, we get 6, 3, 4. These numbers are called the Miller indices of the plane, and the plane is designated as the (634) plane. In (634) plane, h = 6, k = 3 and l = 4.

Similarly the Miller's indices for the plane which the Weiss notation is given by a : 2b : c. Taking reciprocals of coefficients of a, b and c, we get the ratio 1/□, 1/2, 1/1, i.e., 0, 1/2, 1. Multiplying by 2 in order to convert them into whole numbers, we get 0, 1, 2. The plane is designated as the (012) plane in which , h = 0, k = 1 and l = 2.

The distances between parallel planes in a crystal are designated as d_{hkl}. For different cubic lattices these interplanarspacings are given by the general formula

Definition of Miller Indices in three dimensions

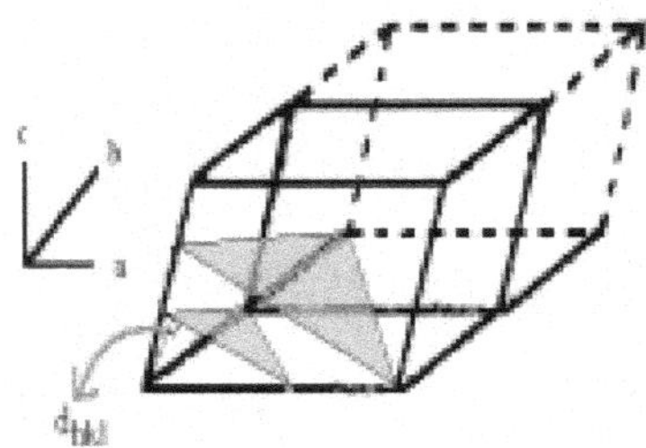

A pair of planes with Miller indices (213)

The **Miller indices** of a face of a crystal are inversely proportional to the intercepts of that face on the various axes.

The **Miller indices** of this particular family of planes are given by the reciprocals of the fractional intercepts along each of the cell directions. e.g. 1/2 x a, 1x b, 1/3 x c.

The procedure for determining the miller indices for a plane is as follows:

1. Prepare a three-column table with the unit cell axes at the tops of the columns.
2. Enter in each column the intercept (expressed as a multiple of a, b or c) of the plane with these axes.
3. Invert all numbers.
4. Clear fractions to obtain h, k and l.

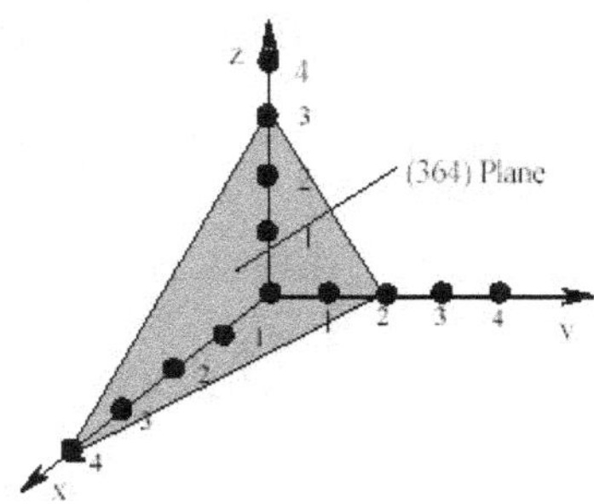

Consider the x-, y-, z- axes in the above figure with the dots representing atoms in a single crystal lattice. To determine the Miller indices, one finds the intercepts on the three axis. The intercepts are: x = 4, y = 2 and z = 3. Then the reciprocals are taken, i.e., ¼, ½, 1/3 and finally these fractions are reduced to the smallest integers, ie., 3, 6, 4 by 12. These are the Miller indices represented as (364)

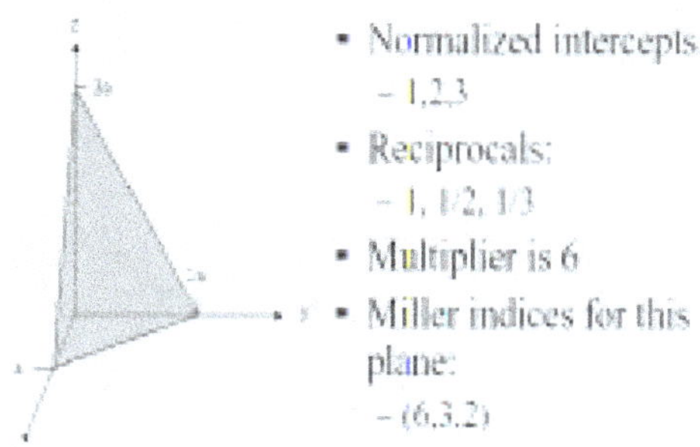

Miller Indices: Example
• Normalized intercepts:
– 1,2,3
• Reciprocals:
– 1, 1/2, 1/3
• Multiplier is 6
• Miller indices for this plane:
– (6,3,2)

Let us look at the most common planes in a cube, shown below in figure. As an example the front crystal face shown here intersects the x-axis but does not intersect the y- or z -axes but parallel to y and z axis. The front crystal face intersects only one of the crystallographic axes(x-axes). So the miller index for the plane is (100). The side plane has intercepts

x = ∞, y = 1, z = ∞ because the plane is parallel to the x- and z-axes, forming the Miller indices gives (010). The top plane has intercepts x = ∞, y = ∞, z = 1 because the plane is parallel to the x- and y- axes, forming the Miller indices gives (001).

The (110) plane intercepts x=1, y=1 and z= ∞ which is parallel to z-axis. Similarly the other two planes are (101) and (011). The (111) plane intercepts all the three axes x=1, y=1 and z=1.

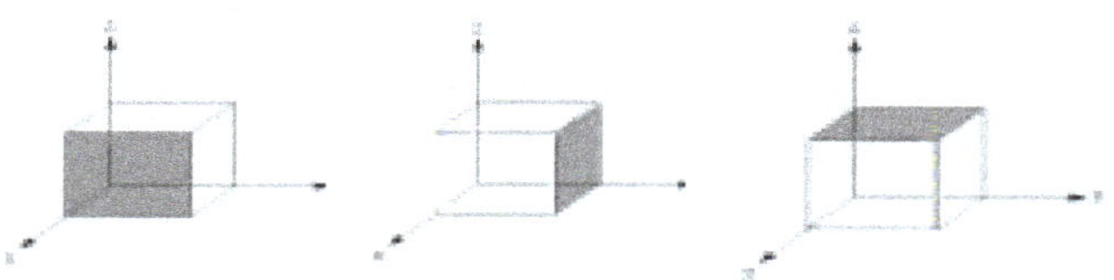

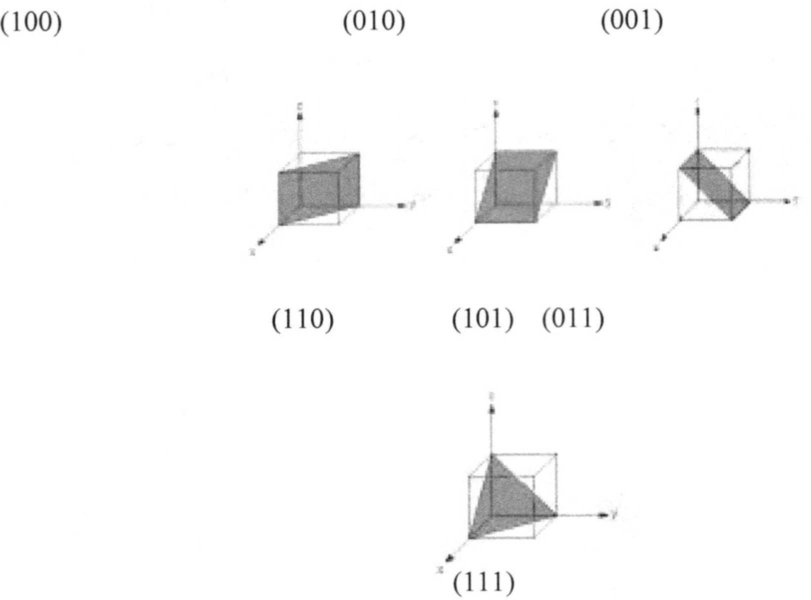

Example 1: Calculate the Miller indices of crystal planes which cutthrough the crystal axes at (i) (2a, 3b, c) (ii) (a, b, c) (iii) (6a, 3b, 3c) and (iv) (2a, -3b, -3c).

Solution:following the procedure given above, we prepare the tables asfollows:

(i) a	b	c	
2	3	1	intercepts
½	1/3	1	reciprocals
3	2	6	clear fractions

Hence, the Miller indices are (326).

(ii) a	b	c	
1	1	1	intercepts
1	1	1	reciprocals
1	1	1	clear fractions

Hence, the Miller indices are (111).

(iii) a	b	c	
6	3	3	intercepts
1/6	1/3	1/3	reciprocals
1	2	2	clear fractions

Hence, the Miller indices are (122).

(iv) a	b	c	
2	-3	-3	intercepts
½	-1/3	-1/3	reciprocals
3	-2	-2	clear fractions

Hence, the Miller indices are (322).

Note: The negative sign in the Miller indices is indicated by placing a bar on the integer. The Miller indices are enclosed within parentheses.

Example 2: How do the spacing of the three planes (100), (110) and(111) of cubic lattice vary?

Applying the formula

$$d_{hkl} = \frac{a}{\sqrt{h^2 + k^2 + l^2}}$$

$$d_{(100)} = \frac{a}{\sqrt{1^2 + 0 + 0}} = a$$

$$d_{(110)} = 1^2 \quad a \quad = a$$

SUMMARY

Solids form an important part of the world around us, providing materials with a definite shape and predictable properties.

Crystalline solids are made of ordered arrays of atoms, ions or molecules.

Amorphous solids have no long-range ordering in their structures.

The unit cell is the basic repeating unit of the arrangement of atoms,ions or molecules in a crystalline solid.

Lattice refers to the three dimensional array of particles in a crystalline solid. Each particle occupies a lattice point in the array.

A simple cubic unit cell has lattice points only at the eight corners of a cube.

A body-centered cubic unit cell has lattice points at the eight corners of a cube and at the center of the cube.

A face-centered cubic unit cell has the same kind of particles (lattice paints) at the eight corners of a cube and at the center of each face.

The geometry of the crystal may be completely defined with the help of coordinate axes meeting at a point.

The miller indices of a face of a crystal are inversely proportional to the intercepts of that face on the various axes.

The study of crystal is known as crystallography.

REFERENCES

1. L.V.Azaroff, Introduction to Solids, TMH edition, Tata MCGraw-Hill, New Delhi.
2. C.Kittel, Introduction to Solid State Physics, Third edition, John Wiley, 1966.
3. A.F.Wells, Structural Inorganic Chemistry, Oxford University Press, 1962.
4. Anthony R. West Solid State Chemistry John Wiley & Sons, New York, 1989.

CHAPTER – 7

GASEOUS STATE

OBJECTIVES

- *To recognize the measurable properties of gases like P,V,T and mass.*
- *To learn various gas laws and ideal gas equation*
- *To learn different units of gas constant `R'.*
- *To understand the Dalton's law of partial pressures & Graham's law of diffusion.*
- *To analyse the deviation of ideal behaviour and to know Vander Waal's equation of state.*
- *To understand and deduce the relationship between critical phenomena and vanderwaal's constants.*
- *To understand Joule - Thomson effect and the role of inversion temperature.*
- *To know methods of liquefaction of gases and adiabatic demagnetization.*

7.1 Properties of gases

Matter is known to exist in three states - solid, liquid and gas. A substance may be made to exist in any one of the three states by varying the temperature or pressure or both. Dynamic motion of molecules therefore is an inherent property in gaseous and liquid states of matter. The energy of motion known as kinetic energy is present in the gaseous molecules. Therefore the basic theory which explains the behaviour of gases is called as kinetic theory of gases.

A gaseous state can be described in terms of four parameters which are known as **measurable properties** such as the volume, V; Pressure, P; Temperature, T and Number of moles, n of the gas in the container.

Pressure effect

A gas may be considered to consist of a large number of molecules moving haphazardly all around in a vessel. Due to their constant motion, the molecules may not collide against one another very frequently, but can strike against the walls of the containing vessel. The molecular collisions are regarded as ideal (ie) perfectly elastic, so that there is no loss of energy in these collisions. **Pressure is defined as force per unit area.** Thisdepends upon the number of molecules that strike per unit area of the walls of the container in one second. The greater the number of molecules striking per unit area of the walls in one second, the greater would be the pressure exerted by the gas. Thus for example, when we pump air into a bicycle tube, the number of molecules within the tube increases and hence the number of collisions of the molecules with the walls per second increases and the pressure goes up.

Temperature effect

The kinetic energy of molecules is given by 1/2 mv^2 where m is the mass of the molecule and v is the velocity of its motion. When a gas is heated, its temperature increases. Although the mass of the molecule remains constant, its velocity increases. This causes an increase in kinetic energy. Therefore the molecules strike the wall of the containing vessel more frequently. In this case there is no change in the number of molecules, but the number of collisions against the walls of the container in a given time increases. Therefore the pressure of the gas increases with rise in temperature when the amount and its volume remain constant.

Volume effect

The volume of the container is considered as the volume of the gas sample. This is considered from the postulates of kinetic theory of gases. That is, the volumes of gas molecules themselves are negligible compared to the container volume. Volume of gas is determined by its pressure, temperature and number of moles at any instant.

Number of moles (n) effect

Effects of pressure and volume of a gas bear a direct proportionality with number of moles. When `n' increases the number of molecules colliding against the wall of container increases. This effect increases the pressure of the gas. When the amount of gas increases the volume occupied by themselves also, increases.

7.2 The gas laws

Boyle's law

Robert Boyle in 1662, studied the effect of change of pressure on the volume of a given mass of gas at constant temperature. According toBoyle's law, for given mass of a gas at constant temperature, the pressure

(P) is inversely proportional to its volume (V).

$$3\ .\frac{1}{V} \quad \text{(at constant temperature)}$$

(or)

$$PV = \text{constant.}$$

Thus if V_1 is the volume occupied by a given mass of a gas at pressure P_1 and V_2 is the volume when pressure changes to P_2, then as the temperature remains constant, according to Boyle's law

$$P_1V_1 = P_2\,V_2 = \text{Constant}$$

Charle's Law

The variation in the volume of a gas with temperature at constant pressure is given by charle's law. The law may be stated as,

For a given mass of gas, at constant pressure, its volume (V) varies directly as its absolute temperature (T).

$$V\ .\ 7 \quad \text{(or)} \quad \frac{V}{T} = \text{Constant}$$

Based on charle's law, the pressure - temperature relation is deduced as, for a given quantity of a gas, at constant volume, the pressure (P) varies directly as its absolute temperature (T)

3 . 7	(or)	$\frac{P}{T}$ = Constant

where T is temperature in kelvin.

The equation of state for an ideal gas

Gases which obey Boyle's law and Charle's law are known as ideal gases. By combining these two laws, an equation of state of an ideal gas can be derived.

According to Boyle's law at constant temperature,

From Charle's law

3 . 7 DW FRQVWDQW YROXPH

By combining these proportionalities,

where `R' is a proportionality constant, commonly called as the **gasconstant.** Generally, the ideal gas equation is written as

$$PV = n\,RT$$

where `n' is the number of moles of the gas.

$$\text{No. of moles} = \frac{\text{Mass of the gas in gram}}{\text{Molecular mass of the gas in gmol}^{-1}} = \frac{m}{M} = \text{mol}$$

$$PV = \frac{M}{M} RT$$

m = mass of the gas.

The ideal gas equation can be written for a constant mass of a gas as,

$$\frac{P_1 V_1}{\quad} = \frac{P_2 V_2}{\quad}$$

T_1 T_2

Standard temperature and Pressure (S.T.P)

The conditions of a gas system present at standard temperature and standard pressure are its temperature at 273K and its pressure being at normal atmospheric pressure namely 1.013×10^5 Nm^{-2} (1 atm). Value of R (Gas constant) depends on the different units of pressure and volume.

7.3 Numerical values of gas constant (R)

The numerical value of the gas constant `R' depends upon the units in which pressure and volume are expressed,

$$R = \frac{PV}{T} \text{ (assuming one mole of gas)}$$

$$R = \frac{\text{Pressure x Volume}}{\text{Temperature}}$$

$$= \frac{\text{Force}}{\text{Area}} \times \frac{\text{Volume}}{\text{Temperature}}$$

Since Volume = Area x length

$$\therefore R = \frac{\text{Force x Length}}{\text{Temperature}}$$

The dimensions of R are thus energy per degree per mole.

a. In litre - atmosphere

One mole of gas at S.T.P occupies a volume of 22.4 litre and thus

At STP

P = 1 atmR = PV/T (for 1 mole)

V = 22.4 litre R = (1 x 22.4)/273

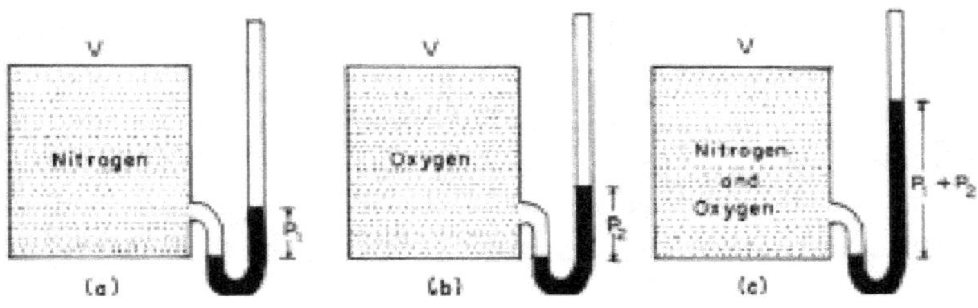

Fig. 7.1 Total pressure eqᵘals the sum of partial pressures of

$T = 273$ K $R = 0.0821$ litreatm $K^{-1}mol^{-1}$

$= 0.0821\ dm^3.atm\ K^{-1}mol^{-1}$

(1 litre = 1 dm^3)

b. In C.G.S. System

At STP 1 mole of gas has

$P = 1$ atm $= 1 \times 76 \times 13.6 \times 980$ dyne cm^{-2}

$= 1.013 \times 10^6$ dyne cm^{-2}

$V = 22400\ cm^3$; $T = 273$ K

$$\therefore R = \frac{PV}{T} = \frac{1.013 \times 10^6 \times 22400}{273}$$

$R = 8.314 \times 10^7$ erg K^{-1} mol^{-1}

c. In M.K.S. System

In MKS or SI units, the unit of R is joule

Since 10^7erg = 1 Joule

$R = 8.314$ Joule K^{-1} mol^{-1}

1 calorie = 4.184 Joule

$R = 1.987$ cals deg^{-1} mol^{-1}

7.4 Dalton's law of partial Pressures

When two or more gases, which do not react chemically, are mixed together in a vessel, the total pressure of the mixture is given by Dalton's law of partial pressures which states that,

"At constant temperature, the total pressure exerted by the gaseous mixture is equal to the sum of the individual pressures which each gas would exert if it occupies the same volume of mixture fully by itself. Partial pressure is the measure of the pressure of an individual gas in a mixture of same volume and temperature.

Thus, if p_1, p_2, p_3 ....are the partial pressures of the various gases present in a mixture, then the total pressure P of the gaseous mixture is given by

$P = p_1 + p_2 + p_3 ...$, provided the volume and temperature of mixture and that of the individual gases are the same.

Equation of state of a Gaseous mixture

Let a gaseous mixture consists of n_A, n_B and n_C moles of three ideal gases A, B and C respectively, subjected to constant T and V, then, according to ideal gas equation.

$$p_A = \frac{n_A RT}{V}; \quad p_B = \frac{n_B RT}{V}$$

And

$$p_c = \frac{n_c RT}{V}$$

Where p_A, p_B, p_C are the partial pressures of A,B,C gases respectively. Hence the total pressure of the mixture is given as

$P = p_A + p_B + p_C$

$PV = (n_A + n_B + n_C) RT$

This equation is known as **equation of state of gaseous mixture.**

Calculation of Partial Pressure

In order to calculate the pressure (p_A) of the individual component say A, in a mixture (A and B), which is equal to the partial pressure of A, according to the equation of state of gaseous mixture it is seen that,

P = Total pressure of the mixture

$$P = (n_A + n_B)\frac{RT}{V}$$

But $p_A = \frac{n_A}{V}RT$ and $p_B = \frac{n_B}{V}RT$

The ratio is given by

x_A = mole fraction of A.

Or $\boxed{p_A = X_A P}$

i.e:- Partial pressure, p_A = mole fraction of A x total pressure.

Similarly;

$p_B = X_B.P$

Thus, the partial pressure of the individual component in the mixture can be calculated by the product of its mole fraction and total pressure.

7.5 Graham's Law of Diffusion

When two gases are placed in contact, they mix spontaneously. This is due to the movement of molecule of one gas into the other gas. This process of mixing of gases by random motion of the molecules is called as **diffusion.**

In 1829, Graham formulated what is now known as Graham's law of diffusion. It states that,

"Under the same conditions of temperature and pressure, the rates of diffusion of different gases are inversely proportional to the square roots of their molecular masses". Mathematically the law can be expressed as

$$\frac{r_1}{r_2} = \sqrt{\frac{M_2}{M_1}}$$

where r_1 and r_2 are the rates of diffusion of gases 1 and 2, while M_1 and M_2 are their molecular masses respectively.

When a gas escapes through a pin-hole into a region of low pressure or vacuum, the process is called **Effusion**. The rate of diffusion of a gas also depends on the molecular mass of the gas. Dalton's law when applied to effusion of a gas is called the Dalton's law of Effusion. It may be expressed mathematically as

$$\frac{\text{Effusion rate of Gas 1}}{\text{Effusion rate of Gas 2}} = \sqrt{\frac{M_2}{M_1}}$$

The determination of rate of effusion is much easier compared to the rate of diffusion. Therefore Dalton's law of effusion is often used to find the molecular mass of a given gas.

7.6 Causes for deviation of real gas from ideal behaviour

The perfect gas equation of state is given by

$$PV = nRT$$

The gases which obey this equation exactly are referred as ideal gases or perfect gases. Real gases do not obey the perfect gas equation exactly. Real gases show deviation because of intermolecular interaction of the gaseous molecules. Repulsive forces between the molecules cause expansion, and attractive forces cause reduction in volume. Under the conditions of low pressure and high

Temperature the inter-molecular interactions of the gaseous

molecules are lower and tend to behave ideally under these conditions. At other conditions of pressure and volume, deviations are seen.

Volume deviation

Based on one of the postulates of the kinetic theory of the gases it is assumed that the volume occupied by the gaseous molecules themselves is negligibly small compared to the total volume of the gas. This postulate holds good for ideal gases and only under normal conditions of temperature and pressure for real gases. When temperature is lowered considerably, the total volume of the real gas decreases tremendously and becomes comparable with the actual volume of gaseous molecules. In such cases, the volume occupied by the gaseous molecules cannot be neglected in comparison with total volume of the gas.

Thus, the volume deviations created at high pressure and low temperature make the real gas to deviate from the ideal behaviour.

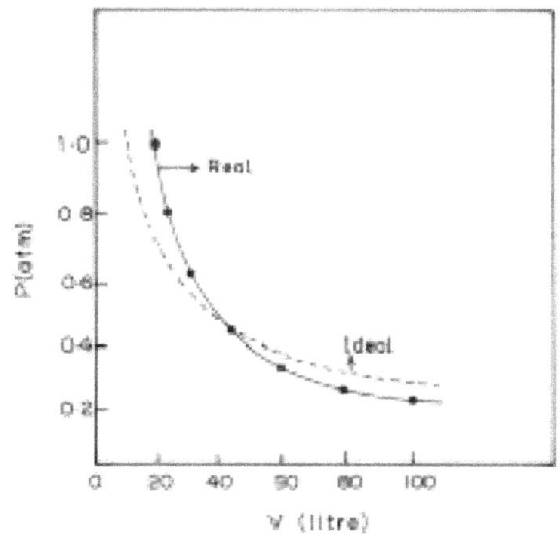

Fig. 7.2 Pressure - volume dependence of ideal and real gases

Pressure deviation

For an ideal gas the forces of attraction between the gaseous molecule are considered to be nil at all temperature and pressure.

For a real gas this assumption is valid only at low pressure or at high temperature. Under these conditions, the volume of the gas is high and the molecules lie far apart from one another. Therefore the intermolecular forces of attraction become negligible.

But at high pressure (or) at low temperature, the volume of the gas is small and molecules lie closer to one another. The intermolecular forces of attraction become appreciable and cannot be neglected. Therefore it is necessary to apply suitable corrections to the pressure of the real gas in the equation of state.

7.7 Vanderwaal's Equation of state

For an ideal gas PV = nRT, is considered as the equation of state. By including the correction terms in the ideal gas equation to account for (i) volume of the gaseous molecules themselves in V and (ii) the intermolecular forces of attraction in pressure, P, the equation of state for the real gas is arrived; J.O Vander Waal's deduced the equation of state of real gases.

i) Volume correction of real gas

The volume of a gas is the free space in the container in which molecules move about. Volume V of an ideal gas is the same as the volume of the container. The volume of a real gas is, therefore, ideal volume minus the volume occupied by the gas molecules themselves. If V_m is the volume of the single molecule then, the excluded volume which is termed as "b" is determined as follows.

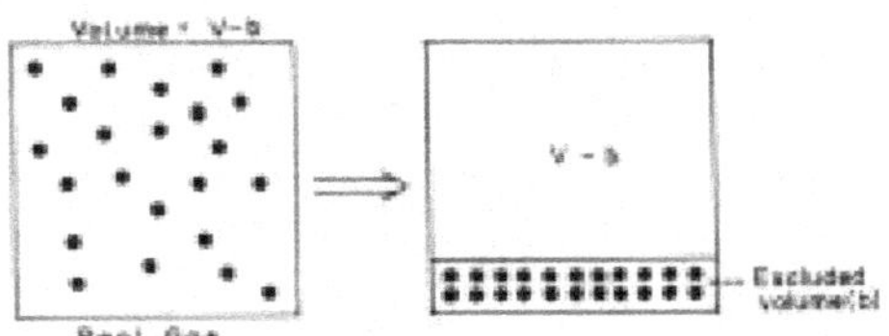

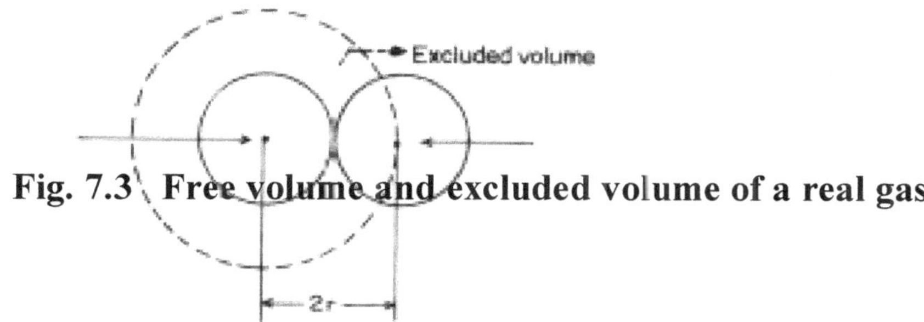

Fig. 7.3 Free volume and excluded volume of a real gas

Fig. 7.4 Collision diameter (2r) and excluded volume of real gas molecules

Let us consider two colliding molecules with radius `r'. The space indicated by dotted sphere of radius 2r will not be available for other molecules to freely move about. (i.e) the dotted spherical volume is known as **excluded volume** per pair of the molecules.

Thus,excluded volume per molecule

$$V_C = \frac{1}{2} \times 8 \; \frac{4}{3}\pi r^3$$

$$= 4\ V_m$$

where, V_m is the actual volume of a single molecule.

The excluded volume for n molecules, 'b' = $4nV_m$, where $4V_m$ is the excluded volume of a molecule.

The corrected volume of the real gas is = (V-b) = free space for molecular movement.

ii) Pressure Correction

In a real gas the pressure deviation is caused by the intermolecular forces of attraction. According to kinetic theory, the pressure of the gas is directly proportional to forces of bombardment of the molecules on the walls of the container. Consider a molecule placed in the interior of the container. It is surrounded equally by other gas molecules in all directions such that the forces of attraction in any direction is cancelled (or) nullified by similar force operating in the opposite direction. However a molecule near the wall of the container which is about to strike is surrounded unequally by other gaseous molecules as shown in Fig 8.5.

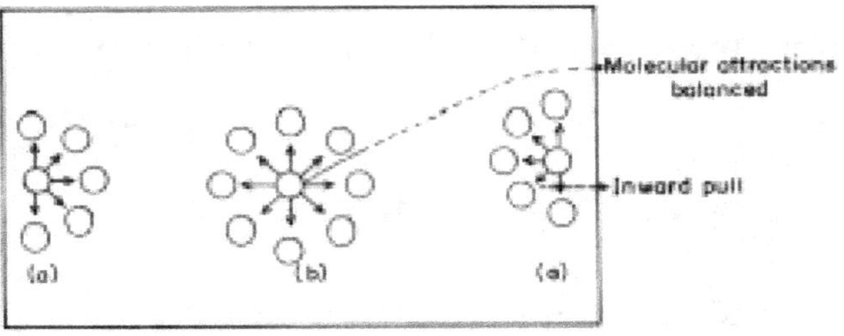

Fig. 8.5 Intermolecular forces of attraction and pressure deviation in real gas molecules

The molecule near the wall experiences attractive forces only such that it will strike the wall with a lower force which will exert a lower pressure than if such attractive forces are not operating on it. Therefore it is necessary to add the pressure correction term to the pressure of the gas to get the ideal pressure. The corrected pressure should be P + p' where p' is the pressure correction factor.

The force of attraction experienced by a molecule near the wall depends upon the number of molecules per unit volume of the bulk of the gas. It is found experimentally that, p' is directly proportional to the square of the density of the gas () which is a measure of the intermolecular attraction of the molecules.

(i.e) $P' \propto \rho^2$ where $\rho = \frac{n}{V}$

Density is inversely related to the volume `V' which is the volume occupied by one mole of the gas. Therefore P' of one mole of the gas is given by

$P' \propto \frac{1}{V^2}$; (or) $P' = \frac{a}{V^2}$

where `a' is a proportionality constant that depends upon the nature of the gas.

Correctedpressure = P + P'

Replacing the term for corrected volume and the corrected pressure in the ideal gas equation for one mole, the equation of state of the real gas is

$$\left(P + \frac{a}{V^2}\right)(V-b) = RT$$

where `a' and `b' are known as Vander Waal's constants.

This equation is also known as Vander Waal's equation of state.

If there are `n' moles of the real gas then the Vander wall's equation becomes

$$\left(P + \frac{n^2a}{V^2}\right)(V-nb) = nRT$$

Units for vanderwaal's constant

The dimensions of the vanderwaal's constant a and b depend upon the units of P and V respectively.

a = atm .dm^6 mol^{-2} (1 litre = 1 dm^3)

Thus a is expressed as atm.dm^6 mol^{-2} units. If volume is expressed in

dm^3 then b is expressed as

Significance of Vander Waal's constant (a) and (b)

1. The term a/V^2 is the measure of the attractive forces of the molecules. It is also called as the cohesion pressure (or) internal pressure.
2. The inversion temperature of a gas can be expressed in terms of \`a' and \`b'

$$T_i = \frac{2a}{Rb}$$

3. The vanderwaal constants \`a' and \`b' enable the calculation of critical constants of a gas.

Limitations of Vander Waal's equation

1. It could not explain the quantitative aspect of deviation satisfactorily as it could explain the qualitative aspects of P and V deviations.
2. The values of \`a' and \`b' are also found to vary with P and T, and such variations are not considered in the derivation of Vanderwaal equation.
3. Critical constants calculated from Vander Waal's equation deviate from the original values determined by other experiments.

7.8 Critical phenomena

The essential condition for the liquefaction of the gas is described by the study of critical temperature, critical pressure and critical volume and their inter relationships.

When a gaseous system is transformed to its liquid state, there is a tremendous decrease in the volume. This decrease in volume can be effectively brought about by lowering of temperature, or by increasing pressure (or) by both. In both these effects the gaseous molecules come closer to each other and experience an increase in force of attraction which results in liquefaction of gases. At any constant temperature when pressure is increased volume is decreased and vice versa. Such P-V curves at constant temperature are known as isotherms. A typical isotherm can be considered similar to Fig.7.2.

The figure 7.2 shows the continuous decrease in pressure with increase in volume for both ideal and real gases. There is a definite deviation exhibited by the real gas from ideal gas behaviour at high pressure and low volumes.

Critical temperature (Tc)

It is defined as the characteristic temperature of a gas at which increase in pressure brings in liquefaction of gas above which no liquefaction occurs although the pressure may be increased many fold. For instance T_c of CO_2 is 31.1°C. This means that it is not possible to liquefy CO_2 by applying pressure when its temperature is above 31.1°C.

Critical pressure (Pc)

It is defined as the minimum pressure required to liquefy 1 mole of a gas present at its critical temperature.

Critical volume (V_c)

The volume occupied by 1 mole of a gas at its critical pressure and at critical temperature is the critical volume (Vc) of the gas.

A gas is said to be at its **critical state** when its pressure, volume and temperature are Pc, Vc and Tc.

7.8.1 Andrews isotherms of carbondioxide

The importance of critical temperature of a gas was first discovered by Andrews in his experiments on pressure-volume isotherms of carbon dioxide gas at a series of temperature. The isotherms of carbon dioxide determined by him at different temperatures are shown in Fig.7.6.

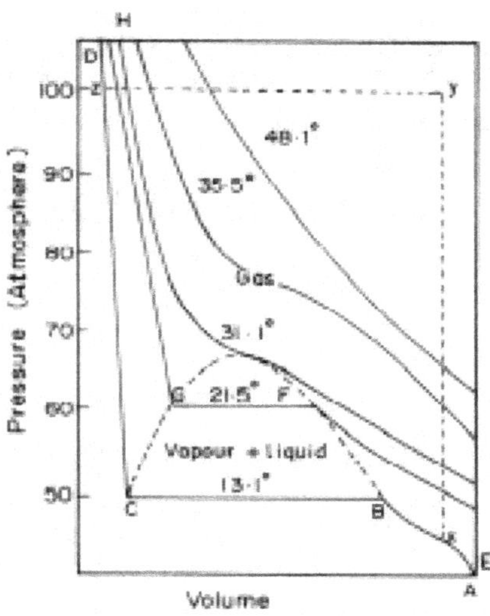

Fig. 7.6 Andrews isotherms of carbondioxide

Consider first the isotherm at the temperature 13.1°C. The point A represents carbon dioxide in the gaseous state occupying a certain volume under a certain pressure. On increasing the pressure its volume diminishes as is indicated by the curve AB. At B, liquefaction of gas commences and thereafter a rapid decrease in the volume takes place at the same pressure, since more and more of the gas is converted into the liquid state. At C, the gas becomes completely liquefied. After `C' the increase of pressure produces only a very small decrease in volume. This is shown by a steep line CD which is almost vertical. Thus, along the curve AB, carbon dioxide exist as gas. Along BC, it exists in equilibrium between gaseous and liquid state. Along CD it exists entirely as a liquid. The isotherm at 21.5°C shows that the decrease in volume becomes smaller because higher the temperature greater is the volume. Therefore more pressure is applied to decrease the volume. This effect makes liquefaction to commence at higher pressure compared to the previous isotherm at 13.1°C.

At still higher temperature, the horizontal portion of the curve becomes shorter and shorter until at 31.1°C it reduces to a point. The temperature 31.1°C is regarded as the critical temperature of CO_2. At this temperature, the gas passes into liquid imperceptibly. Above 31.1°C the isotherm is continuous. CO_2 cannot be liquefied above 31.1°C no matter how high the pressure may be. The portion of area covered by curve H with zyx portion always represents the gaseous state of CO_2.

7.8.2 Continuity of state

Thomson's experiment

Thomson (1871) studied the isotherm of CO_2 drawn by Andrews. He suggested that there should be no sharp points in the isotherms below the critical temperature. These isotherms should really exhibit a complete continuity of state from gas to liquid. This, he showed as a theoretical wavy curve. The curve MLB in Fig.7.7 represents a gas compressed in a way that it would remain stable. The curve MNC represents a superheated liquid because compression above T_c, leads to heating effects. This type of continuity of state is predicted by Vander Waal's equation of state which is algebraically a cubic equation. The Vander Waal's equation may be written asMultiplying by V^2

$$PV^3 - (RT + Pb) V^2 + aV - ab = 0$$

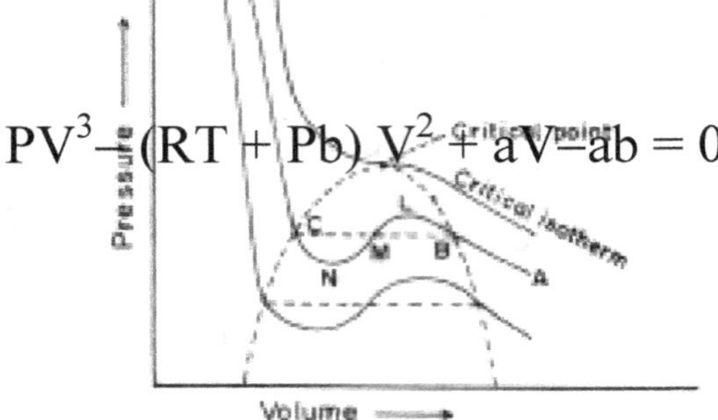

Fig. 7.7 Thomson's isotherms of CO_2

Thus, for any given values of P and T there should be three values of V. These values are indicated by points B,M and C of the wavy curve. The three values of V become closer as the horizontal part of the isotherm rises. At the critical point the three roots of Vanderwaal 'V' become identical and there is no longer any distinction between the gas and liquid states. Here, the gas is said to be in critical state. This effect enables the calculation of Tc, Pc and Vc in terms of Vander Waal's constants.

7.8.3 Derivation of critical constants from Vanderwaal's constants

Let us derive the values of critical constants T_c (critical temperature), V_c (critical volume) and P_c (critical pressure) in terms of the Vanderwaal's constants `a' and `b'. The Vanderwaal's equation is given by

Rearranging this equation in the powers of V

$$V^3 - \left(\frac{RT}{P} + b\right)V^2 + \frac{aV}{P} - \frac{ab}{P} = 0 \quad (4)$$

For this cubic equation of V, three roots (values of V) are possible. At the critical point, the three values of V become identical and is equal to the critical volume (V_c).

Therefore $V = V_c$ at T_c

$\therefore (V - V_c) = 0 \quad (5)$

$\therefore (V - V_c)^3 = 0 \quad (6)$

Upon expanding this equation

$$V^3 - 3V_cV^2 + 3V_c^2V - V_c^3 = 0 \quad (7)$$

This equation is identical with the cubic equation derived from Vander Waal's equation if we substitute T by T_c and P by P_c.

7.9 Joule-Thomson Effect

Joule-Thomson showed that when a compressed gas is forced through a porous plug into a region of low pressure, there is appreciable cooling.

The phenomenon of producing lowering of temperature when a gas is made to expand adiabatically from a region of high pressure into a region of low pressure, is known as Joule-Thomson effect.

When the gas is allowed to escape into a region of low pressure the molecules move apart rapidly against the intermolecular attractive forces. In this case work is done by the gas molecules at the expense of internal energy of the gas. Therefore cooling occurs as the gas expands. This reduction in the temperature is generally referred as Joule-Thomson effect and is used in the liquefaction of gases.

7.10 Inversion temperature (Ti)

The Joule-Thomson effect is obeyed by a gaseous system only when its temperature is below a characteristic value. The characteristic temperature below which a gas expands adiabatically into a region of low pressure through a porous plug with a fall in temperature is called as inversion temperature (Ti).

Ti is characteristic of a gas and it is related to the Vanderwaal's constant `a' and `b',

At the inversion temperature there is no Joule Thomson effect (ie) there is neither fall nor rise in temperature. Only when the temperature of the gas is below the inversion temperature there is a fall in temperature during adiabatic expansion. If the temperature of the gas is above Ti there is a small rise in temperature. For gases like H_2 and He whose Ti values are very low -80°C and -240°C respectively, these gases get warmed up instead of getting cooled during the Joule-Thomson experiment. These gases will obey Joule-Thomson effect only when they are cooled to a temperature below these Ti values.

7.11 Conditions of liquefaction of gases

Many industrial processes require large quantities of liquid air, liquid ammonia, liquid carbondioxide etc. The production of liquids from various gases is therefore an important commercial operation.

There are different methods of liquefaction of gases, such as (i) based on the concept of critical temperature followed by the compression (ii) based on Joule-Thomson effect (iii) Adiabatic demagnetization.

In the case of gases like NH_3, Cl_2, SO_2 and CO_2 whose Tc values are near and below the ordinary temperatures, they can be liquefied easily by increasing the pressure alone at their respective Tc values.

Gases like H_2, O_2, N_2 and He have very low T_c values and hence Joule Thomson effect may be applied to bring in effective cooling.

Helium is cooled by Joule-Thomson effect to a lower temperature and further cooling for its liquefaction, is carried out by the method of adiabatic demagnetisation.

Linde's Method

This method makes use of Joule Thomson effect and is used to liquefy air or any other gas. Pure air or any gas is first compressed to about 200 atmospheres and is allowed to enter the innertube of the concentric pipes as shown in Fig.7.8. The valve v of jet J is then opened and the gas is allowed to expand suddenly into the wider chamber C.

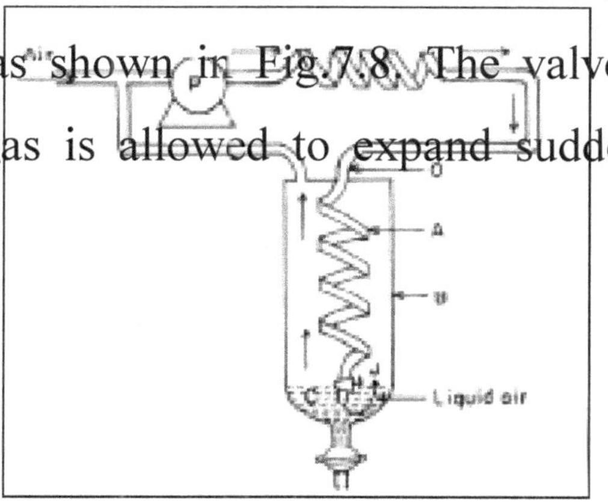

Fig. 7.8 Linde's apparatus for liquefaction of gas

The gas gets cooled due to expansion and its pressure is reduced to about 50 atm. The gas is now allowed to pass through the outer tube `O'. At this stage the incoming gas is initially cooled by the outgoing gas. Further cooling of the incoming gas occurs during expansion in the chamber C. The cooled gas is again compressed and is circulated in. By repeating the process of cooling and compression followed by expansion, the gas is liquefied and finally the liquid air drops out from the jet into the bottom of chamber C.

Claude's process

In this method compressed air is allowed to do mechanical work of expansion. This work is done at the expense of the kinetic energy of the gas and hence a fall of temperature is noted. This principle is combined with Joule-Thomson effect and utilised in Claude's process of liquefaction of air. Air is compressed to about 200 atmospheres and is passed through the pipe ABC (Fig.7.9). At C, a part of the air goes down the spiral towards the jet nozzle J and a part of the air is led into the cylinder D provided with an air tight piston. Here the air moves the piston outwards and expands in volume as a result of which considerable cooling is produced. The cooled air passes up the liquefying chamber during which process it cools the portion of the incoming compressed air. The precooled incoming compressed air then experiences Joule-Thomson expansion when passed through Jet nozzle J and gets cooled further. The above process takes place

repeatedly till the air is liquefied

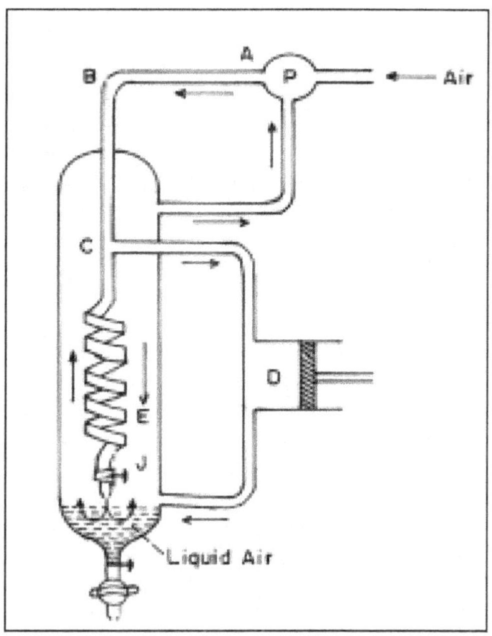

Fig. 7.9 Claude's apparatus for liquefaction of air

Adiabatic demagnetization

Generally, the method used to reach the very low temperature of about 10^{-4} K is adiabatic demagnetization. In this method the paramagnetic samples such as Gadolinium sulphate is placed surrounding the gas sample and cooled to about 1K along with the gas in any one of the cooling methods. The paramagnetic sample used in this method is suddenly magnetized by the application of strong magnetic field. This magnetization (ordering of molecular magnets) occurs while the sample surrounds the cooled gas and has thermal contact with the walls of the container. When the magnetic field is suddenly removed, demagnetization occurs which brings in a disordered state of the molecular magnets. To reach this state thermal energy is taken away from the cooled air such that its temperature gets further lowered. By this technique, as low as zero kelvin can be reached.

SUMMARY

- P,V,T and mass are the measurable properties of gas. They obey Boyle's and Charle's law. The equation of state for an ideal gas in PV = nRT.
- For constant mass of a gas,

$$\frac{P_1V_1}{T_1} = \frac{P_2V_2}{T_2}$$

- Different units of R :0.0821 it atm $K^{-1}mol^{-1}$;
  8.314 x 10^7 erg K^{-1} mol^{-1}; 8.314 joule K^{-1} mol^{-1}; 1.987 cal K^{-1} mol^{-1}.
- Equation of state of gaseous mixture is $PV = (n_A + n_B + n_C)$ RT.
- By Graham's law, (diffusion rate1/diffusion rate2) = $(M_2/M_1)^{1/2}$ (or) (effusion $rate_1$/ effusion $rate_2$) = $(M_2/M_1)^{1/2}$
- Real gases deviate from V_{ideal} and P_{ideal}. The equation of state of real gas = Vanderwaal equation.
- Critical temperature, critical pressure, critical volume represent the critical state of the gas. Andrew's isotherm describes critical temperature of carbon dioxide. Thomson's experiment describes continuity of state.

P_c, V_c, T_c are related to Vander Waal's constants a and b as V_c = 3b;

$$P_c = \frac{a}{27b^2}; \quad T_c = \frac{8a}{27Rb}$$

- Joule Thomson effect predicts adiabatic expansion of a compressed gas through an orifice to cause a fall in temperature. Inversion temperature = $2a/R_b$ is the temperature below which Joule Thomson effect is obeyed.
- Liquefaction of gases is carried out by Linde's and Claude's processes adopting Joule-Thomson effect. Liquefaction of Helium and zero kelvin are achieved by adopting adiabatic demagnetization.

REFERENCES:

Text book of physical chemistry, Lewis and Glasstone.

www.ingramcontent.com/pod-product-compliance
Lightning Source LLC
Chambersburg PA
CBHW080810180526
45168CB00006B/2386

* 9 7 8 1 5 1 7 2 2 0 1 3 6 *